Science Versus Textbook Evolution

Mace Baker, Ph.D.

New Century Books
Lakeland, Florida

About the Author

Mace Baker has taught young people for over 20 years in the public school system. He earned his B.A. from Biola University, and his Ph.D. from Pacific University in Missouri. He has been engaged in Creation Research for more than 20 years and has been conducting seminars in America, Canada, Western and Eastern Europe, Russia, the Philippines, India, and Central America.

Cover design by Darryl Hodson

Acknowledgments:

Many thanks to Ed and Charlene Rosa and to Darryl Hodson for their very helpful preliminary proof-reading of this volume, and I am especially indebted to Melissa Pagnucco whose excellent secretarial work and editing has helped bring this volume to completion. This has included assistance in typing, proof-reading, and formatting. Her carefulness and attention to detail has been accompanied by a wonderful attitude, which has made each work session a positive experience.

Printed and bound in the United States of America

For information on Dr. Mace Baker's lectures and seminars, the author may be contacted at macebaker@myyellowstone.net

ISBN 0-9647283-3-8

Introduction

This book is written for the purpose of making it possible for parents of students in the public schools, and for the public at large, to be aware of the evolutionary propaganda which is present in all of the state-adopted high school biology texts. It is also written for the purpose of showing the tremendous amount of scientific evidence that is antagonistic to all of this evolutionary propaganda. It will be noted that the objective evidence in the world of nature is clearly antagonistic to the evolution model, but not to the Creation model, as presented in the book of Genesis.

It is important for me to be very clear about one aspect of this book. The vast majority of quotes, which are antagonistic to the theory of evolution, are actually quotes from evolutionists. Probably on only one or two occasions, do I quote from an individual, such as Michael Denton, who is technically not an evolutionist. He and his kind have been associated with the title "non-creationists." This means that he is neither a Christian nor a creationist, yet he has written a book showing the flaws in the theory of evolution.

It is also important that I emphasize this point very forthrightly--I am not by any means implying that the evolutionists, which I quote, have changed their mind about the value of evolution and have become sympathetic to the idea of creation. Each one that I quote, as best as I can ascertain, is indeed a loyal evolutionist and is totally committed to the theory of evolution. They are simply stating accurate facts of nature with respect to the particular issue at hand, and though the information they relate is unsupportive of this theory, they themselves would readily confess that they firmly believe in the consensus--that is, in the theory of evolution.

Finally, let me provide a quick and simple explanation of how the book reads. Generally speaking, each category will begin with a brief introduction. This will be followed by a subtitle, Textbook Evolution. In this section we will reference the actual statements from the high school textbooks. Then, this will be followed by a second subtitle, Science Versus. In this section we will provide a large amount of evidence that is contrary to the statements made in the high school textbooks. These statements are being made by university level teachers and researchers, as well as writers of magazines such as *Science*, *Scientific American*, and *Nature*, all of which are supportive of the theory of evolution. Each major chapter ends with a summary.

FOSSIL RECORD

Introduction

In all of these state-adopted high school biology textbooks, there is in the section showing the evidence for evolution, a chapter devoted to the fossil record. In these pages, the students are informed that the fossil record shows the history of life over a long period of time. Further, the students are led to believe that there is good evidence in the fossil record, by means of transitional forms, showing that evolution over time has taken place. In this chapter, we will review these assertions from a variety of textbooks and, by contrast, show the enormous amount of evidence against these assertions.

In addition to the textbooks, a variety of research scientists have emphasized the importance of the fossil record to the theory of evolution. Dunbar of Yale, in his college text, states,

> "Although the comparative study of living plants and animals may give very convincing circumstantial evidence, fossils provide the only historical, documentary evidence that life has evolved from simpler to more and more complex forms."[1]

Pierre P. Grasse, of the University of Sorbonne in France, remarks,

> "Naturalists must remember that the process of evolution is revealed only through fossil forms. A knowledge of paleontology is, therefore a prerequisite; only paleontology can provide them with the evidence of evolution and reveal its course or mechanisms."[2]

Steven M. Stanley, of Johns Hopkins University, also emphasizes the fossil record's importance to evolution,

> "While many inferences about evolution are derived from living organisms, we must look to the fossil record for the ultimate documentation of large-scale change. In the absence of a fossil record, the credibility of evolutionists would be severely weakened. We might wonder whether the doctrine of evolution would qualify as anything more than an outrageous hypothesis.[3]

The Holt, Rinehart, & Winston high school textbook, *Biology: Principles and Explorations*, like all of the other high school textbooks, is very emphatic about the importance of the fossils to evolution,

> "The most direct evidence that macro evolution has occurred comes from fossils."[4]

One of the most glaring errors of the evolutionists' view of how fossils are formed can be seen from this statement from one of the high school life science texts (*California Focus on Life Science*),

> "Most fossils form when organisms that die become buried in sediments. Sediments are particles of soil and rock. When a river flows into a lake or ocean, the sediments that the river carries settle to the bottom. Layers of sediments may cover the dead organisms. Over millions of years, the layers may harden to become sedimentary rock."[5]

This assumption of the slow and gradual means by which sedimentary rocks form in the ocean as a result of river transportation, was also promoted by Charles Lyell, the father of Uniformitarianism. Before the end of the 19th century, the Challenger Expedition demonstrated emphatically that sedimentary rocks are not forming by this means at all. In 1872, the British government equipped the Challenger with dredging equipment. Then, for the next four years they went about examining the bottoms of all the oceans. However, in all that time searching they were unable to find any stratified rock in the process of becoming a sedimentary rock formation. As a result, the late geologist, Sir Archibald Geikie, acknowledges that,

> "Thanks to the great work done by the Challenger another national expedition, we have learned that the accumulations now forming on the deeper parts of the ocean floor have no analogy among the formations of the earth's crust. They differ indeed entirely from any other formation which geologists have considered to be of deep sea origin."[6]

Further, Philipp H. Kuenan states in his book on marine geology,

> "...there are no indications in stratigraphy that true ocean bottoms have ever been transformed into continental masses."[7]

The work of the Challenger and the results obtained should have brought an end to the idea of uniformitarianism. However, as we will see more and more in this volume, the main and most effective means by which the concept of uniformitarianism and evolution are promoted are not the facts of science but consensus--that is, the theory and world view that most scientists in the university system prefer.

Textbook Evolution

Regarding transitional types, the textbooks make these general statements,

> "Although the fossil record is incomplete, it does provide evidence that evolution has indeed occurred."[8] (Alton Biggs, Chris Kapicka, Linda Lundgren, *Biology: The Dynamics of Life*, Glencoe, McMillian/McGraw-Hill)
>
> "When the *Origin of Species* was published, no intermediates had been found....But there are now many good examples of evolutionary transitions."[9] (George B. Johnson, *Biology:Visualizing Life*, Holt, Rinehart & Winston)
>
> "For the last two centuries, paleontologists have studied fossils in the earth's strata (layers) all over the world and have pieced together the story of past life. The fossil record is rich in information. One of its most striking patterns is the succession of life forms over time."[10] (Sylvia S. Mader, *Biology,* McGraw Hill)
>
> "Fossils offer the most direct evidence that evolution takes place." Then on the same page these authors also state, "...Darwin predicted that intermediate forms between the great groups of organisms would eventually be found. Since Darwin's time, many of these intermediaries have been found. For example, fossil intermediaries have been found between fishes and amphibians,

between reptiles and birds, and between reptiles and mammals, adding valuable evidence about the fossil history of the vertebrates."[11] (George B. Johnson and Peter H. Raven, *Biology*, Holt, Rinehart and Winston, Inc.)

"Fossils are an important source of evolutionary evidence because they provide a record of early life and evolutionary history."[12] (Alton Biggs, et al., *Biology*: *The Dynamics of Life*, Glencoe)

"Researchers have discovered many hundreds of transitional fossils that document various intermediate stages in the evolution of modern species from organisms that are now extinct."[13] (K.R. Miller, J. Levine, *Biology,* Prentice Hall)

"Fossils are among the most powerful evidence of evolution."[14] (John H. Postlethwait & Janet L. Hopson, *Modern Biology*, Holt, Rinehart, & Winston)

"Fossils are an important source of evolutionary evidence because they provide a record of early life and evolutionary history."[15] (Alton Biggs, et. al., *Biology: The Dynamics of Life*, Glencoe Science, *National Geographic*, McGraw-Hill/Glencoe)

"Though Darwin recognized the limitations of the fossil record, he predicted the existence of fossils intermediate in form between species. Today, scientists studying evolutionary relationships have found hundreds of thousands of transitional fossils that contain features shared by different species."[16] (Alton Biggs, Whitney Crispen Hagins, William G. Holliday, Chris L. Kapicka, Linda Lundgren, Ann Haley MacKenzie, William D. Rogers, Marion B. Sewer, Dinah Zike, *Biology,* Glencoe)

Now, let us examine the textbook's remarks in particular categories.

Bacteria

Textbook Evolution

Note how in the Holt 1994 text the students are introduced to fact and to assumption in such a way that leads the student to believe that the assumption is also a fact.

> "When you have a case of strep throat, your cells are under attack by bacteria. Bacteria are single-celled organisms. Bacterial cells are much less complicated than your cells. They are also the ancestors of your cells. Over the course of time, bacterial cells evolved. The result was the kinds of cells that make up your body."[17] (George B. Johnson, *Biology Visualizing Life*, Holt, Rinehart & Winston)

The first two sentences of this paragraph are factual and verifiable; however, the last two sentences of this paragraph are assumptions and are not verifiable.

Farther down in the page they provide another verifiable fact,

> "All cells can be divided into two large categories. A eukaryotic cell, or **eukaryote**... is a large complex cell that contains a membrane-bound compartment called a **nucleus.** The nucleus contains DNA within chromosomes. A prokaryotic cell, or **prokaryote**... is a small, simple cell that lacks a nucleus. Its DNA is a single, circular molecule that is not enclosed in a membrane-bound compartment."[18] (*Ibid.*)

Then they follow up with an assumption in the guise of a fact,

> "Fossils of the first cells on Earth, which existed about 3.5 billion years ago, reveal that these ancient cells were prokaryotes."[19] (*Ibid.*)

These different kind of bacteria cells have actually been found as fossils. This is a fact. That they existed 3.5 billion years ago is not a fact. It is an assumption that is part of the evolution model. The sad thing is that the high-school students assume, since these statements are in their textbook, that they are all equally true, equally verifiable. If someone took the time to teach them the difference between a fact and an assumption they would soon see that the theory of evolution is made up of primarily of assumptions and not facts.

At the top of page 56, entitled, "Inside the Cell," there are four objectives listed for the student. The fourth objective says, "Describe how eukaryotes evolved from prokaryotes."[20] (*Ibid.*)

There is actually no information on this page by which a student could obtain adequate information for such a description.

In the paragraph below, which is found in the Prentice Hall textbook, we have four sentences. The first, second, and fourth sentences are assumptions based on the theory of evolution and are by no means verifiable. The third sentence, referring to "soft-bodied" organisms, we will address later.

> "Aerobic forms of life evolved, and eukaryotes appeared. Some of these organisms gave rise to multicellular forms that continued to increase in complexity. Few fossils exist from this time because the animals were all soft-bodied. Life existed only in the sea."[21]
> (Kenneth R. Miller & Joseph Levine, *Biology*, Prentice Hall)

Science Versus

According to the theory of evolution, things evolve from the simple to the complex. However, with regard to many of the fossil organisms of the Cambrian layer, they are already very complex and there are no decreasingly complex organisms of the same kind in lower rocks. McShey, an evolutionary biologist at the University of Michigan, states,

> "Everybody knows that organisms...get more complex as they evolve. The only trouble with what everyone knows is that there is no evidence it's true."[22]

Furthermore, as Steven Stanley of the University of Hawaii at Manoa, writes, there is no fossil evidence supporting the evolution of Precambrian one-celled organisms to the large variety of complex multicellular marine organisms of the Cambrian period.

> "The evidence is now mounting that most of the major fossil groups of the Cambrian arose by rapid evolution.... In the first place, fossil assemblages consisting of the imprints of soft-bodied creatures ...have been found in many areas of the world, but are never older than latest Precambrian."[23]

Preston Cloud long ago reported that,

> "There are as yet no unequivocal metozoa in rocks of indisputable proterozoic or older age."[24]

Andrew Knoll of Harvard University states,

> "Many of the prokaryotes from Spitzbergen and related areas exhibit characteristics of morphology, development and behavior (as inferred from their orientation in the sediments) that render them virtually indistinguishable from cynanobacteria and other bacteria that live in the comparable habitats today... Some Spitzbergen eukaryotes resemble modern prasinophyte (green) algae, whereas others bear closer resemblance to the so-called chromophytes algae such as the dinoflagellates that are ubiquitous in modern oceans."[25]

In other words, in the world's "oldest" rocks, we find that the prokaryotes and eukaryotes have existed from the beginning. They are still with us today, "billions of years later," and yet they are essentially unchanged in their form and structure. The fact is that we have two types of bacteria cells on the Earth. The assumption is that one preceded the other, and the students are obligated to accept this as a fact, as per "Describe how eukaryotes evolved from prokaryotes."[26]

The Holt text tells the students confidently that there are primitive cells. For instance,

"Primitive prokaryotes dominated the early Precambrian, and 2 billion years passed before eukaryotic organisms with nuclei, appear in the fossil record."[27]

However, as French biologist, Jacques Monod, reports,

"...we have no idea what the structure of a primitive cell might have been. The simplest living system known to us, the bacterial cell...in...its overall chemical plan is the same as that of all other living beings. It employs the same genetic code and the same mechanism of translation as do, for example, human cells. Thus the simplest cells available to us for study have nothing 'primitive' about them...no vestiges of truly primitive structures are discernible."[28]

Evolution, in the high school textbooks, is defined as change through time or descent with modification. For example,

"The fossil record provides evidence about the history of life on Earth. It also shows how different groups of organisms have changed over time."[29]

"However, evolution is descent with modification."[30]

It is interesting that the consensus of opinion is coming to the conclusion that the most obvious characteristic of the fossil record is stasis, that is to say, no change. One of the most important displays of this phenomenon has to do with the life-form which we know as algae. Paleontologists have, according to the evolutionary timetable, found algae and bacteria fossils which theoretically date back 3.4 billion years. That's a very long time. In an article in *Science News* we have a question and the answer,

"How do those ancient life forms, say, the algae, compare to those that exist today? Evidence from paleontology suggests that the algae haven't changed much."[31]

The research further shows,

> "Even more intriguing, the pleurocapsalean algae turned out to be almost identical to modern pleurocapsalean algae at the family and even at the generic level."[32]

In fact, Steven M. Stanley comments,

> "Once established, an average species of animal or plant will not change enough to be regarded as a new species, even after surviving for something like a hundred thousand, or a million, or even ten million generations....something tends to prevent the wholesale restructuring of a species, once it has become well established on earth."[33]

Doesn't this, then, underscore the improbability that "established" invertebrates would not be apt to change to a fish, and that any "established" fish species would not be apt to change to an amphibian, etc. and etc.?

Even one of the strongest evolutionary spokesmen of our day, Richard Dawkins, acknowledges that,

> "Some species of the unjustly called 'primitive' amoebas have as much information in their DNA as 1,000 Encyclopedia Britannicas."[34]

A paper presented to *Proceedings of the National Academy of Sciences* presents information in regard to "...the evolution of eukaryotes (cells with nuclei) from prokaryotes (cells without nuclei)...."[35]

As a consequence of this research, the case for the evolution of eukaryotes (cells with nuclei) from prokaryotes (cells without nuclei) is not at all established,

> "Do the Harvard paleontologists' findings shed any light on the origin of eukaryotes (cells with nuclei) from prokaryotes (cells without nuclei): Probably not."[36]

In other words, the idea that prokaryotes evolved into eukaryotes is simply a matter of evolutionary lore, and not a matter of observable facts of nature,

> "In brief, as Barghoorn puts it, 'We have no really good evidence from all of the Precambrian records... of a genuine eukaryotic cell.'"[37]

The account of Creation, as recorded in the book of Genesis, is relegated to a creation myth, yet the history of evolution, as espoused in the textbook, is devoid of fossil documentation. Instead, what is documented by the fossils is what is found in the book of Genesis, that the Creator created the animals separate from one another, and created them to reproduce 'after their kind'. Stefan Bengtson is an evolutionist who considers the Genesis account of Creation to be a myth, but he has to acknowledge that the facts of science harmonize very nicely with the Genesis account,

> "If any event in life's history resembles man's creation myths, it is this sudden diversification of marine life when multi-cellular organisms took over as the dominant actors in ecology and evolution. Baffling (and embarrassing) to Darwin, this event still dazzles us and stands as a major biological revolution on par with the invention of self-replication and the origin of the eukaryotic cell. The animal phyla emerged out of the Precambrian mists with most of the attributes of their modern descendants."[38]

Precambrian/Cambrian

Introduction

According to the evolutionary time table, the oldest rocks are those referred to as Precambrian. For a long time, it appeared that there were no fossilized life forms to be found in these rocks. This was a perplexing problem for the theory of evolution because the layer of rocks above the Precambrian, known, of course, as the Cambrian rocks, contains innumerable and highly complex life forms, such as crabs, lobsters, and trilobites. Eventually, a variety of fossilized, tiny life forms (bacteria) were found to be in the Precambrian rocks. This discovery, however, has proved to be an insurmountable problem for the theory of evolution because the time zones of the Precambrian and Cambrian represent hundreds of millions of years. In those hundreds of millions of years, there should be found a very large amount of fossilized organisms, which represent the transitional species, which were to be the ancestors of the complex Cambrian fauna. However, no such organisms have ever been found. As a result, the writers of textbooks promoting evolution must keep using the word "abrupt," meaning that the Cambrian organisms appear in the fossil record abruptly and that there are no transitional types which illustrate a sequence of ancestors bridging the gap from the tiny bacteria forms to the complex organisms of the Cambrian rocks.

Textbook Evolution

In the Glencoe text students are told of the Precambrian Era,

"The first 4 m of the geologic time ribbon in Figure 14.5 make up the Precambrian. This is nearly 90 percent of Earth's entire history, stretching from the formation of Earth to the beginning of the Paleozoic era about 542 million years ago. The Precambrian was an important time. Earth formed and life first appeared. Eventually, autotrophic prokaryotes, much like the cyanobacteria that made the stromatolites in Figure 14.6, enriched the atmosphere with oxygen. Eukaryotic cells also emerged, and by the end of the Precambrian, life was flourishing, and the first animals had appeared."[39] (Alton Biggs, *et. al*, *Biology*, Glencoe)

Then they are told of the Paleozoic Era,

"A drastic change in the history of animal life on Earth marked the start of the Paleozoic...era. In the space of just a few million years, the ancestors of most major animal groups diversified in what scientists call the Cambrian explosion. Not all major groups of organisms evolved rapidly at this time, and paleontologists still do not know when the rapid changes started or ended."[40] (*Ibid.*)

Other textbooks inform the students with similar information,

"All of today's groups of animals can trace their ancestry to this time and perhaps earlier according to new molecular clock data."[41] (Sylvia S. Mader, *Biology*, McGraw Hill)

"Geologic time begins with the Precambrian...Time. Although few multicellular fossils exist from this time, the Precambrian actually covers about 88 percent of Earth's history...."[42] (Kenneth R. Miller & Joseph Levine, *Biology*, Pearson/Prentice Hall)

Science Versus

Students are not told two very disturbing facts with regard to the Pre-Cambrian/Cambrian eras. First, they are not told that the so-called Cambrian era with its many complex organisms appears abruptly or catastrophically as opposed to the theorized slow and gradual scenario. However, the late and very popular Harvard scientist, Stephen Jay Gould, recognizes this phenomenon,

"The Cambrian Explosion occurred in a geological moment, and we have reason to think that all major anatomical designs may have made their evolutionary appearance at that time. ...not only the phylum Chordata itself, but also all its major divisions, arose within the Cambrian Explosion. So much for chordate uniqueness... Contrary to Darwin's expectation that new data would reveal gradualistic continuity with slow and steady expansion, all major discoveries of the past century have only heightened the massiveness and geological abruptness of this formative event..."[43]

The second disturbing fact that the students are not told is that there are no known ancestors in the earlier Precambrian Era for all of these "many types of invertebrates" as well as other creatures found in the Cambrian Era, even though this has been well documented,

Premier dinosaurologist, Edwin H. Colbert, and his colleague, Marshall Kay, admit to the absence of evolutionary ancestors for the complex organisms found in the Cambrian layer,

"The introduction of a variety of organisms in the early Cambrian, including such complex forms of the arthropods as the trilobites, is surprising...The introduction of abundant organisms in the record would not be so surprising if they were simple. Why should such complex organic forms be in rocks about six hundred million years old and be absent or unrecognized in the records of the preceding two billion years? ...If there has been evolution of life, the absence of the requisite fossils in the rocks older than the Cambrian is puzzling."[44]

Kay and Colbert asked this question back in the 1960's. In the 1980's Stephen Jay Gould is still puzzled,

"Studies that began in the early 1950s and continue at an accelerating pace today have revealed an extensive Precambrian fossil record, but the problem of the Cambrian explosion has not receded, since our more extensive labor has still failed to identify any creature that might serve as a plausible immediate ancestor for the Cambrian fauna... Where, then are all the Precambrian ancestors--or, if they didn't exist in recognizable form, how did complexity get off to such a fast start?"[45]

Darwin himself understood the problem that the Cambrian explosion poses for the theory of evolution. Stephen Gould writes, in *The Panda's Thumb*,

"The fossil record had caused Darwin more grief than joy. Nothing distressed him more than the Cambrian explosion, the coincident appearance of almost all complex organic designs..."[46]

Darwin's own words are,

"The case at present must remain inexplicable; and may be truly urged as a valid argument against the views here entertained."[47]

The British geologist, Thomas Neville George, also acknowledges,

"Granted an evolutionary origin of the main groups of animals, and not an act of special creation, the absence of any record whatsoever of a single member of any phyla in the Pre-Cambrian rocks remains as inexplicable on orthodox grounds as it was to Darwin."[48]

Eugene Koonin, an expert in the field of evolutionary and computational biology reveals,

"The Cambrian explosion in animal evolution during which all the diverse body plans appear to have emerged almost in a geological instant is a highly publicized enigma. Although molecular clock analysis has been invoked to propose that the Cambrian explosion is an artifact of the fossil record whereas the actual divergence occurred much earlier, the reliability of these estimates appears to be questionable. In an already familiar pattern, the relationship between the animal phyla remains controversial and elusive."[49]

Science writer, Rihcard Monastersky, asserts,

"Before the Cambrian period, almost all life was microscopic, except from some enigmatic soft-bodied organisms. At the start of the Cambrian, about 544 million years ago, animals burst forth in a rash of evolutionary activity never since equaled. Ocean creatures acquired the ability to grow hard shells, and a broad range of new body plans emerged within the geologically short span of 10 million years. Paleontologists have proposed many theories to explain this revolution but have agreed on none."[50]

For the evolutionist, this, as Gould states, is a gigantic problem,

"The Cambrian explosion was the most remarkable and puzzling event in the history of life."[51]

So, we see all of these science writers and science researchers freely admitting to the same amazing problem. The lowest geological layer with significant fossils is the Cambrian. It is bursting with millions of fossils and many of them represent very complex organisms. However, in the layer below, the Pre-Cambrian, there are essentially, for all intents and purposes, no organisms that could qualify as ancestors of the Cambrian animals. This so called problem should, for the objective reader, illustrate how unscientific their uniform approach is. They are unwilling to entertain the suggestion that the Pre-Cambrian rocks do not contain the ancestors of the Cambrian organisms because the Pre-Cambrian organisms did not have evolutionary ancestors.

The English scientist, Richard Dawkins, who was interviewed in the movie, *Expelled*, at least admits that this great problem is favorable to the Creationist position,

> "Eldredge and Gould certainly would agree that some very important gaps really are due to imperfections in the fossil record. Very big gaps, too. For example the Cambrian strata of rocks, vintage about 600 million years, are the oldest ones in which we find most of the major invertebrate groups. And we find many of them already in an advanced state of evolution, the very first time they appear. It is as though they were just planted there, without any evolutionary history. Needless to say, this appearance of sudden planting has delighted creationists."[52]

The amazing transition between Pre-Cambrian life forms and the extremely complex Cambrian life forms has also been acknowledged in a major article by J. Madeleine Nash, a senior science correspondent for *TIME* magazine. The ten to fifteen million years they refer to is strictly theoretical and is based upon uniformitarian principles. As can be seen from a variety of evidences in the sedimentary rocks, and especially in the fossilization of many animals' soft parts (as listed below), the rocks, on a global scale, support catastrophism rather than uniformitarianism,

> "Then, 543 million years ago, in the early Cambrian, within the span of no more than 10 million years, creatures with teeth and tentacles and claws and jaws materialized with the suddenness of apparitions. In a burst of creativity like nothing before or since, nature appears to have sketched out the blueprints for virtually the whole of the animal kingdom. This explosion of biological diversity is described by scientists as biology's Big Bang."[53]

Richard Dawkins, author of *The Blind Watchmaker*, also recognizes the problem that the organisms of the Cambrian rocks are embarrassingly advanced,

> "...the Cambrian strata of rocks, vintage about 600 million years, are the oldest in which we find most of the major invertebrate groups. And we find many of them already in an advanced state of evolution, the very first time they appear. It is as though they were just planted there, without any evolutionary history."[54]

According to the theory of evolution, things evolve from the simple to the complex. However, with regard to many of the fossil organisms of the Cambrian layer, they are already very complex and there are no decreasingly complex organisms of the same kind in lower rocks. In an article in *Discover* magazine, Lori Oliwenstein, writes,

> "Everybody knows that organisms...get more complex as they evolve. The only trouble with what everyone knows, says McShea, an evolutionary biologist at the University of Michigan, is that there is no evidence it's true."[55]

The excuse given by some evolutionists for the lack of transitional species from Precambrian protozoa to the vast amount of creatures in the Cambrian Era, is that the Precambrian creatures lacked hard parts and therefore did not fossilize well. This excuse is undermined by the fact that none of the Precambrian fossils had hard parts, including the famous Ediacaran fauna, (yet they were fossilized) and that many of the fossilized Cambrian creatures did not have hard parts (such as the jellyfish). As paleontologists have come to learn, many animals in the past have been buried so quickly that not only their bones, but also their soft parts have been fossilized. Derek E. G. Briggs, the director of the Yale Peabody Museum of Natural History, reports,

> "In certain circumstances the organic tissues themselves can survive for geologically significant periods of time."[56]

In general, the public conceives of the fossil record as including exclusively fossilized animals containing hard parts; however,

> "Mineralized skeletons are not ubiquitous in the animal kingdom. Two-thirds of existing phyla lack any mineralized hard parts."[57]

By this Briggs means that two-thirds of existing phyla are devoid of hard parts that could have been fossilized in the past. This means that many of our modern animals have not been found in the fossil record, not because they didn't exist in previous eras, but because, due to their anatomical make-up, were not easily fossilized. He also acknowledges,

> "The discovery in Waukesha, Wisconsin, of the first significant assemblage of soft-bodied animals from the Silurian period, 400 million years ago, extends the known time range of some taxa back several millions of years."[58]

David M. Martill of the University of Portsmouth tells us that,

> "An examination of the large collections of Santana Formation fossil fishes....suggests that soft-tissue fossilization in this formation is relatively common. In some fish specimens nearly the entire musculature of the trunk may be preserved, and in most of the three-dimensional specimens there is some phosphatization of soft tissues."[59]

To make matters even worse, paleontologist, Dale A. Russell of the North Carolina State University, informs us that we even have fossilized fish droppings,

> "The bodies of fish were often abundantly preserved on the stagnant lake floors, scattered among fish droppings measuring up to 15 centimeters long."[60]

Finally, also in Brazil, we have not only bodily parts of pterosaurs fossilized, but bacteria as well. Science writer, Carl Zimmer, states,

> "...when an animal died in the Araripe water, it was quickly coated with a layer of sediment. How quickly the nodules formed is hard to say, but judging from the presence of fossilized muscle and sometimes even bacteria on the skin of Araripe animals, it could have happened within hours."[61]

If the soft parts of these animals (and even fish droppings) could be fossilized, then, in the theorized vast extent of the Precambrian Era, surely the theoretical ancestors of the Cambrian fauna could have been fossilized. After all, the Glencoe high school text reports that,

> "The Precambrian Era accounts for about 87 percent of Earth's history...."[62]

Contrary to popular opinion, the fossil record is not at all poor with regard to sites in which the animals fossilized had soft parts. Derek Briggs informs us that,

> "From the beginning of the Cambrian the number of known sites displaying significant soft-part preservation exceeds 60 and for each of these major sites there are many minor ones."[63]

The textbooks tell the students of the fossilized fauna which characterized the Precambrian Era and the fossilized fauna which characterized the Cambrian Era, but are unable to explain to the students the immense problem connected with the Cambrian explosion. Simply put, all major phyla explode into existence in the Cambrian Era, but in the vast expanse of the Precambrian Era, there are no ancestors for the Cambrian fauna. This problem is discussed by Valentine and Erwin (James W. Valentine is with the University of Chicago and Douglas H. Erwin is the senior scientist and Curator of paleobiology at the Smithsonian's National Museum of Natural History),

> "If ever we were to expect to find ancestors to or intermediates between higher taxa, it would be in the rocks of the late Precambrian to Ordovician times, when the bulk of the world's higher animal taxa evolved. Yet transitional alliances are unknown or unconfirmed for any of the phyla or classes appearing then."[64]

Further, in an article in *American Scientist*, Valentine and Cathryn A. Campbell also confirm this perplexing problem for evolutionists,

> "The abrupt appearance of higher taxa in the fossil record has been a perennial puzzle. Not only do characteristic and distinctive remains of phyla appear suddenly, without known ancestors, but several classes of phylum, orders of a class, and so on, commonly appear at approximately the same time, without known intermediates."[65]

Brachiopoda are among the most common invertebrate fossils found in abundance, yet Professor Derek Ager of the University of Swansea, Wales, admits that he has been unable to trace any of them back to their theoretical ancestors,

> "It must be significant that nearly all the evolutionary stories I learned as a student, from Trueman's *Ostrea/Gryphea* to Carruther's *Raphrentis delanouei*, have now been 'debunked.' Similarily, my own experience of more than twenty years looking

for evolutionary lineages among the Mesozoic Brachiopoda has proved them equally elusive."[66]

Author H.S. Ladd comments,

"Most paleontologists today give little thought to fossiliferous rocks older than the Cambrian, thus ignoring the most important missing link of all. Indeed the missing Pre-Cambrian record cannot properly be described as a link for it is in reality, about nine-tenths of the chain of life; the first nine-tenths."[67]

In their book, *Zoology*, Hickman, Roberts, and Larson make the following four statements,

"Unraveling the origin of the muticellular animals (metazoans) has presented many problems for zoologists,"[68]

"The origin of the cnidarians and ctenophores (comb jellies) is obscure,"[69]

"The relationship of the crustaceans to other arthropods has long been a puzzle,"[70]

"Despite the excellent fossil record, the origin and early evolution of the echinoderms (sea stars) are still obscure,"[71]

Finally, it is very important to notice the fact that must be very embarrassing to evolution theorists. Supposedly, life evolved from the simple to the complex over millions of years of time, but the fossil record shows that all phyla appeared at the very beginning. Monastersky, in an article in *Science News*, explains,

"The remarkably complex forms of animals we see today suddenly appeared....This moment, right at the start of the Earth's Cambrian Period...marks the evolutionary explosion that filled the seas with the earth's first complex creatures...'This is Genesis material,' gushed one researcher....demonstrates that the large animal phyla of today were present already in the early Cambrian..."[72]

Daniel Axelrod of the University of Michigan acknowledges that fish also appear in the Cambrian rocks,

> "This report of fish material from Upper Cambrian rocks further extends the record of the vertebrates by approximately 40 million years."[73]

Monastersky goes on to say,

> (Distinct fossil kinds) "Demonstrates that the large animal phyla of today were present already in the early Cambrian and that they were as distinct from each other as they are today...a menagerie of clam cousins, sponges, segmented worms, and other invertebrates that would seem vaguely familiar to any scuba diver."[74]

Though it is not a useful scientific fact for the theory of evolution, it has become abundantly clear that all of the major animal phyla are present in the "oldest" fossil-bearing rocks, the Cambrian layer. As Douglas Futuyma, the distinguished professor of ecology and evolution at Stony Brook University, reveals,

> "It is considered likely that all the animal Phyla became distinct before or during the Cambrian, for they all appear fully formed, without intermediates connecting one form to another."[75]

Distinguished professor in the Department of Ecology at Yale University, Jeffrey S. Levinton, concurs,

> "The phyla do seem to have appeared suddenly and simultaneously. For that reason, some paleontologists refer to the Cambrian 'explosion.'"[76]

Is it not interesting to note that the book of Genesis states that all of the major animal groups were created in a very short period of time and that they were essentially contemporaries? The research above certainly does not indicate that the fossil record provides any evidence to the contrary.

Invertebrates to Fish

Textbook Evolution

According to the evolutionary textbooks, the evolution from invertebrates to fishes went through an "intermediate group."

> "Scientists think that vertebrates evolved from urochordates. Remember from Chapter 30 that Urochordata is a chordate subphylum that includes tunicates. Most urochordates are sessile as adults but have a free-swimming larval stage. Based on the characteristics of modern tunicate larva, scientists theorize that the ancestor of vertebrates was a free-swimming urochordate that lacked a sessile adult stage."[77] (Bernice Essenfeld, Carol Gontang, Randy Moore, *Biology*, Addison-Wesley)

> "The earliest vertebrates were jawless fishes with bony skeletons."[78] (George B. Johnson, *Biology Visualizing Life*, Holt, Rinehart & Winston)

The Addison-Wesley text shows a chart on page 552B showing that modern fishes are divided into three different categories: Agnatha (jawless fishes) Osteichthyes (cartilaginous fishes), and Chrondrichthyes (bony fishes).

> Under the title "The Natural History of Fishes," the Addison Wesley states for its objective for this chapter, "Describe the evolution of the three modern classes of fishes." Then they list for the teacher under the section called Teaching Strategies, the following, "The earliest fishes were jawless and covered with bony plates"..."Cartiliginous and bony fishes probably evolved separately from armor-covered, jawed fishes"..."Evidence shows that bony fishes first evolved in fresh water and later spread out into the oceans."[79] (Bernice Essenfeld, Carol Gontang, Randy Moore, *Biology*, Addison-Wesley)

> "Fishes were the first vertebrates to evolve. They did not arise directly from tunicates or lancelets, but fishes and nonvertebrate chordates probably did evolve from common invertebrate ancestors."[80] (Kenneth R. Miller & Joseph Levine, *Biology*, Pearson/Prentice Hall)

"Jawless fishes evolved. Jawless fishes were the first vertebrates."[81] (Michael Padilla, et al., *California Focus on Life Science*, Pearson/ Prentice Hall)

Science Versus

First, Francis Hitching explains that the expected intermediates between invertebrates and vertebrates are missing.

> "Museums have, for instance, countless piles of fossils of the early invertebrate sea creatures, and an equally large number of ancient fishes. Between the two, covering a period of about 100 million years, there ought to be cabinets full of intermediates -- indeed, one would expect the fossils to blend so gently into one another that it would be difficult to tell where the invertebrates ended and vertebrates began.
>
> "But this isn't the case. Instead, groups of well-defined, easily classifiable fish jump into the fossil record seemingly from nowhere: mysteriously, suddenly, full-formed, and in a most un-Darwinian way. And before them are maddening, illogical gaps where their ancestors should be."[82]

Above, the Addison-Wesley text, *Biology*, indicates to the students that creatures such as tunicates might suffice as a transitional link between invertebrate and fish, because they have cartilaginous skeletons. Actually, this only creates more problems for the theory of evolution as F.D. Ommanney, author of the *Life Nature Library: The Fishes*, explains:

> "How this earliest chordate stock evolved, what stages it went through to eventually give rise to truly fish-like creatures we do not know...there is a gap of perhaps 100 million years we will probably never be able to fill."[83]

Alfred S. Romer, American paleontologist and specialist in vertebrate evolution, admits that finding ancestors for the chordates is unlikely,

> "We have no certain fossil record of lower chordates or chordate ancestors and very possibly never shall have."[84]

Hickman, Roberts, and Larson further comment,

> "However, the exact phylogenetic position of the chordates within the animal kingdom is unclear,"[85]

Professor Mustafa Kuru, a Turkish evolutionary biologist is confident that chordates evolved from invertebrates but acknowledges that there are no transitional forms,

> "There is no doubt that chordates evolved from invertebrates. However, the lack of transitional forms between invertebrates and chordates causes people to put forward many assumptions."[86]

The theory of evolution is not at all helped by the observation that essentially all invertebrates began at the very beginning, which, in evolutionary terms, refers to Precambrian or Cambrian sediments. Theoretically, only the most primitive life forms should have arisen then. David Raup, paleontologist at the University of Chicago, explains,

> "Nearly all living phyla of marine invertebrates that have reasonably good fossil records have first occurrences either in the late Precambrian or early to middle Cambrian. At the class level there are 27 paleontologically important living groups and all have documented occurrences which are Silurian or older."[87]

Supposedly, jawed fishes arose from jawless fishes, but as Romer comments,

> "A persisting major gap in our paleontological record, however, is the almost complete absence of any trace of an earlier jawed fish."[88]

Regarding the mollusks, there are apparently unbridgeable gaps. Bruce Runnegar and John Pojeta Jr. (adjunct scientist at the Smithsonian) state,

> "Stasek theorized that the extant mollusks are the progeny of three separate lineages that separated before the phylum was well established. He wrote that no known intermediate forms, fossil or

living, bridge the 'enormous gaps between any two of the three lineages,' and therefore treated each as a separate subphylum."[89]

Writing in *Paleobiology*, Hopson and Radinksy also reference the problem of the ancestry of jawed fishes,

"No known group of agnathans can be considered as lying close to the ancestry of jawed fishes."[90]

This problem is also acknowledged by vertebrate paleontologist, Robert Carroll,

"We still have no evidence of the nature of the transition between cephalochordates and craniates. The earliest adequately known vertebrates already exhibit all the definitive features of craniates that we can expect to have preserved in fossils. No fossils are known that document the origin of jawed vertebrates."[91]

Next, with regard to the origin of the bony fishes, evolutionary paleontologist, Gerald T. Todd, shows that we find more questions than answers in the fossil record,

"All three subdivisions of the bony fishes first appear in the fossil record at approximately the same time. They are already widely divergent morphologically, and they are heavily armored. How did they originate? What allowed them to diverge so widely? How did they all come to have heavy armor? And why is there no trace of earlier, intermediate forms?"[92]

Surely one of the most reasonable answers to the last question Dr. Todd asked is that there were no earlier intermediate forms.

The high school textbook states,

"Arthropods most likely evolved from the annelids. As arthropods evolved, body segments became reduced in number...."[93]

On the following page, p 771, there is a colorful diagram/picture, showing the 'radiation' of the classes of arthropods,

"**Figure 31.9** The radiation of classes of arthropods on the Geologic Time Scale shows their relationships. Which group is largest? Which are the oldest living arthropods?"[94]

In order to answer the question, "Which are the oldest living arthropods?" the students are obligated to review the diagram and come up with the answer, "Merostomates." This is because, in the diagram, Merostomates are shown as originating in the Precambrian, while the other arthropods are shown as originating in the Paleozoic.

Colin Tudge, British science writer, is not enthusiastic about these opinions, with regard to the origin of the arthropods,

"All of which zoological folklore, is... a load of rubbish... One of the fundamental weaknesses of the traditional view of arthropod phylogeny--that all forms arose from the polychaestes via a common stock--is that it presupposed the existence of intermediate ancestral types. But there is no fossil evidence for such types and there are no modern animals resembling those hypothetical types."[95]

With regard to the origin of fishes in general, experts such as paleontologist Barbara Stahl, who do the field research state the following:

"The higher fishes, when they appear in the Devonian period, have already acquired the characteristics that identify them as belonging to one or another of the major assemblages of bony or cartilaginous form."[96]

The late E.H. Colbert, previously a curator at the American Museum of Natural History, along with co-authors, Morales, & Minkoff, also highlight the lack of fossil evidence supporting the evolution of fish,

"Both these groups [bony and cartilaginous] appeared in the late Silurian period, and it is possible that they may have originated at some earlier time, although there is no fossil evidence to prove this."[97]

J.O. Long, in his book, *The Rise of Fishes: 500 Million Years of Evolution*, further acknowledges this problem,

"Yet the transition from spineless invertebrates to the first backboned fishes is still shrouded in mystery, and many theories abound as to how the changes took place."[98]

With regard to the evolution of sharks, Dr. John Long of the Museum Victoria, Melbourne, Australia comments:

"The mystery remains as to how sharks first evolved... The current fossil evidence is too incomplete to answer this question."[99]

He goes on to say,

"The origin of sharks are still a mystery."[100]

Concerning the evolution of the spiny fin fish, Dr. Long relates that,

"The structure of the head and the shape of the body in the earliest complete canthodian [spiny fin] fossils tell us little about which other group of fishes they may have evolved from or collaterally with."[101]

Actually, the entire evolutionary history of all fishes is devoid of fossil evidence, as Errol White, a Fellow of the Royal Society (F.R.S.), tells of the total lack of fossil evidence supporting the idea that fish evolved from invertebrates,

"But whatever ideas authorities may have on the subject, the Lung-fishes, like every other major group of fishes that I know, have their origins firmly based in nothing....[102]

The evolution theory insists that invertebrates evolved into fish. However, as C.P. Hickman and his co-authors, Roberts and Larson, indicate in their book, *Integrated Principles of Zoology*, that there is no fossil record to justify this confidence,

"The very first fishes undoubtedly arose from invertebrate protochordates...However, the first fishes left no fossil record and their form and relationships are a mystery."[103]

J.R. Norman of the Department of Zoology at the British Museum of Natural History agrees,

"The geological record has so far provided no evidence as to the origin of the fishes..."[104]

Fish to Amphibian

Textbook Evolution

The students are told that fishes were the first vertebrates to evolve, that this evolution began 500 million years ago, and that there are fossils to support these two "facts,"

"The oldest vertebrate fossils, dating back to about 500 million years ago, are of fishes similar to the one shown in Figure 31.1."[105] (George B. Johnson & Peter H. Raven, *Biology: Principles and Explorations*, Holt, Rinehart and Winston, Inc.)

They also state, that there was an "age" of fish that lasted many millions of years before any higher forms of life evolved,

"For more than 100 million years, fishes were the only vertebrates."[106] (*Ibid.*)

Then the students are told that amphibians evolved from some of these fishes,

"More than one-half of all living vertebrates are fishes. The largest and most diverse vertebrate group, fishes are the evolutionary base from which amphibians, the first land vertebrates, arose."[107] (*Ibid.*, p. 745)

"The significance of fin-to-limb evolution is critical to the understanding of vertebrate evolution."[108] (*Ibid.*, p. 744)

The Holt textbooks are emphatic that amphibians evolved from fishes, and in particular from lobe-finned fishes,

"The first vertebrates able to live on land were amphibians. This group of animals first appeared about 370 million years ago. Amphibians evolved from lobe-finned fishes, a group of bony fishes that includes coelacanths and lungfishes."[109] (*Ibid.*, p. 706)

The Glencoe textbook states the same thing,

"Scientists believe that either the lungfishes or lobe-finned fishes were directly ancestral to amphibians, the first land vertebrates."[110] (*Biology, The Dynamics of Life*, Glencoe, McMillian/McGraw-Hill)

The textbooks tell the students that amphibians evolved from lobe-finned fishes around 350 to 390 million years ago,

"Amphibians first appeared about 350 million years ago. It is widely believed that amphibians evolved from lobe-finned fishes, *Figure 33.15*, around the middle of the Paleozoic Era."[111] (*Ibid.*, p. 830)

"Fossil evidence indicates that amphibians evolved from lobe-finned fishes about 390 million years ago."[112] (G.B. Johnson and P.H. Raven, *Biology: Principles & Exploration*, Holt, Rinehart, & Winston, Inc.)

"Fossil evidence points to the lobe-finned fish as the ancestor of modern amphibians."[113] (Bernice Essenfeld, Carol Gontang, Randy Moore, *Biology*, Addison-Wesley)

It is theorized that regular fins evolved into lobe fins so that the fish could eventually crawl on the land,

"Lobed fins probably arose as structures for moving on bottoms of ponds. These fishes could use the lobed fins to crawl across land to another pond..."[114] (*Ibid.*)

In 1996, Johnson and Raven writing in the Holt textbook, *Biology: Principles & Exploration*, identified the Ichthyostega as one of the first amphibians to evolve from fish,

> "Ichthyostega and other early amphibians were the first vertebrates to live largely on land, beginning about 370 million years ago."[115]

However, in 1998, *National Geographic* admits that the amphibian which evolved from a fish is still unknown,

> "They are the closest living relatives of that earliest, though still unknown, amphibian."[116] (Hans Fricke, "Coelacanths: The Fish that Time Forgot" *National Geographic* Vol. 173, No. 6)

> "About 370 million years ago, the first amphibians evolved from lobe-finned bony fishes and became the first vertebrates to live on land."[117] (*Modern Biology*, Holt, Rinehart, & Winston)

> "The first amphibians to climb onto land probably resembled lobe-finned fishes similar to the modern coelacanth. However, the amphibians had legs...."[118] (Kenneth R. Miller and Joseph Levine, *Biology*, Pearson/Prentice Hall)

Science Versus

However, a wide variety of university professors and researchers seem to manifest less confidence to how this miraculous change came about, than we find stated in the high school text.

According to Barbara Stahl, none of today's fishes qualify as being ancestors of the amphibians,

> "...none of the known fishes is thought to be directly ancestral to the earliest land vertebrates. Most of them lived after the first amphibians appeared, and those that came before show no evidence of developing the stout limbs and ribs that characterized the primitive tetrapods..."[119]

> "Since the fossil material provides no evidence of other aspects of the transformation from fish to tetrapod, paleontologists have had to speculate how legs and aerial breathing evolved..."[120]

Actually, as Gordon R. Taylor states in his book, *The Great Evolution Mystery*, the simple truth is that there are no fossils to substantiate the textbook statements regarding the evolution of fish to amphibian,

> "There are no intermediate forms between finned and limbed creatures in the fossil collections of the world."[121]

Hopson and Radinsky say the same thing,

> "No clearly intermediate form in the fish-tetrapod transition has been discovered."[122]

Robert Carroll agrees,

> "We have no intermediate fossils between rhipidistian fish and early amphibians...."[123]

The coelacanths have been regarded as having lived about 70 million years ago, and, since no fossils of them were seen in more recent rocks, it was thought that they had become extinct. However, fishermen in 1938 actually hauled up a living coelacanth. They were able to identify it because it looked so much like the fossils in the museum. This means that in approximately 70 million years (evolutionary time) the fish had not changed one iota. Now, since they were known to be alive and well, *National Geographic* decided to send divers down to observe these fish. They, of course, hoped to be able to video them walking about on the bottom of the sea. This would be strong evidence for them that, over time, the fins could have evolved into legs. Pascale Chenel briefly describes them in his book, *Life and Death of Dinosaurs*,

> "This extraordinary fish was a coelacanth, representative of a group considered extinct since the end of the Cretaceous. Since then, some fifty of these fabulous fish have been caught; they are about 1.5 meters (5 feet) long and weigh some 70 kilograms (320 pounds)."[124]

As mentioned above, hope was strong that they would find the fish using their lobe fins to walk about on the sea bottom. However, after being successful in locating the fish, studying and photographing them, *National Geographic* reports the following,

> "Though we observed several individuals resting with their fins braced against the sea bottom, we never saw any of them walk, and it appears the fish is unable to do so."[125]

Surely, with regard to the idea that coelacanths evolved into amphibians, these recent observations should make it abundantly clear that no such imaginary scenario ever happened.

Now, with regard to the origin of the three orders of amphibians, Robert Carroll elucidates,

> "Despite these similarities, frogs, salamanders and caecilians are very different from one another in skeletal structure and ways of life, both now and throughout their known fossil record...we have found no fossil evidence of any possible antecedents that possessed the specialized features common to all three modern orders....In the absence of fossil evidence that frogs, salamanders, and caecilians evolved from a close common ancestor, we must consider the possibility that each of the modern orders evolved from a distinct group of Paleozoic amphibians."[126]

Colbert and Morales, reiterate the same facts in different words,

> "Despite these similarities, there is no evidence of any Paleozoic amphibians combining the characteristics that would be expected in a single common ancestor. The oldest known frogs, salamanders and caecilians are very similar to their living descendants."[127]

Years ago there was a lot of talk about the ichthyostegid amphibians being the descendants of fish. However, as Hopson and Radinsky share,

> "The oldest known tetrapods, the ichthyostegid amphibians of the Late Devonian, though first reported on in 1932 and represented by numerous specimens, have never been completely described. No

clearly intermediate form in the fish-tetrapod transition has been discovered...."[128]

In conclusion, Malcolm W. Browne's candid comments in an article for *New York Times* about the fish to amphibian transition should be in the high school textbook, along with the party-line propaganda,

> "No one can be certain which group or groups of fishes was the first to make the transition to land, or what their evolutionary pathways may have been...the transition from water to land occurred so long ago, and various family trees suggested by the fossil record are so tangled that scientists acknowledge they may never be able to sort them out definitively."[129]

Amphibian to Reptile

Textbook Evolution

According to the state adopted high school biology text, the evolutionists are confident that reptiles evolved from amphibians. However, as can be seen from the two different texts below, they differ in their confidence as to the exact date.

> (Figure 34.10) "Reptiles evolved from amphibians 300 million years ago during the late Paleozoic Era."[130] (Alton Biggs, Chris Kapicka, Linda Lundgren, *Biology, The Dynamics of Life*, Glencoe, McMillian/ McGraw-Hill)

> "From studies of fossils and comparative anatomy, biologists infer that reptiles arose from amphibians."[131] (*Modern Biology,* Holt, Rinehart, & Winston)

> "Reptiles evolved from amphibian ancestors about 340 million years ago."[132] (George B. Johnson, Peter H. Raven, *Biology*, Holt, Rinehart and Winston)

> "To colonize dry habitats, animals needed a way to reproduce that did not require depositing eggs into water. Reptiles, which evolved from

amphibian-like ancestors, were the first vertebrates to develop this adaptation."[133] (Kenneth R. Miller & Joseph Levine, *Biology*, Pearson/ Prentice Hall)

Science Versus

Note the presence of the word "unfortunately" in Robert Carroll's remarks with reference to this theorized evolutionary process,

> "Unfortunately not a single specimen of an appropriate reptilian ancestor is known prior to the appearance of true reptiles. The absence of such ancestral forms leaves many problems of the amphibian-reptilian transition unanswered."[134]

Snakes

Here again is another "unfortunately" with regard to evidence for evolution in this information provided by Colbert and Morales,

> "Unfortunately, the fossil history of the snakes is very fragmentary, so that it is necessary to infer much of their evolution from the comparative anatomy of modern forms."[135]

However, as Dunbar of Yale (previously cited) comments, the use of comparative anatomy provides no "historical documentary evidence,"

> "Although the comparative study of living plants and animals may give very convincing circumstantial evidence, fossils provide the only historical, documentary evidence that life has evolved from simpler to more and more complex forms."[136]

With regard to snakes, paleontologist Michael J. Benton writes,

> "...are believed to have arisen from 'lizard' ancestors, but the nature of those ancestors is a mystery."[137]

Turtles

There are no amphibian ancestors for turtles in the fossil record. The first or oldest turtles are easily recognized as turtles--true turtles, not partial turtles. As Colbert states,

> "The first true turtles made their appearance by the late part of the Triassic period, by which time they were far advanced along the lines of adaptive radiation typical of modern turtles."[138]

Dinosaurs

Textbook Evolution

> "Dinosaurs are reptiles. They evolved from the crocodile-like reptiles known as thecodonts...."[139] (George B. Johnson and Peter H. Raven, *Biology: Principles & Exploration*, Holt, Rinehart & Winston)

> "On land, the most famous and spectacular reptiles, the dinosaurs, appeared and evolved into a great variety of forms during the Mesozoic era."[140] (*Modern Biology*, Holt, Rinehart, & Winston)

> "Therapsids, reptiles with complex teeth and legs positioned beneath their body, gave rise to mammals about the same time dinosaurs evolved, during the Triassic period."[141] (George B. Johnson, Peter H. Raven, *Biology*, Holt, Rinehart and Winston)

> "Beginning about 235 million years ago, dinosaurs dominated life on land for roughly 150 million years. They evolved from the *thecodonts*, an extinct group of crocodile-like reptiles."[142] (Alton Biggs, et. al., *Biology: The Dynamics of Life*, Glencoe Science, *National Geographic*, McGraw-Hill/Glencoe)

Science Versus

Dinosaurs are divided into two main orders: the saurischians, or lizard hip dinosaurs, and the ornithischians, or bird hip dinosaurs. The ornithischians include such varied kinds of dinosaurs as the ceratopsians (horned dinosaurs), the hadrosaurs, the ankylosaurs (armored dinosaurs), and the ornithopoda consisting of such species as the iguanodontids. Of this entire order, dinosaurologist, Dale Russell says,

> "The origin of the numerous and interesting groups of ornithischian dinosaurs is unclear."[143]

Lessem and Glut, in their *Dinosaur Encyclopedia*, comment,

> "As far as we know, the ornithischians (bird-hipped dinosaurs) originated sometime during the Triassic Period, but we do not know the details of their early evolution. The group's origin lies in as-yet-undiscovered fossils at the base of the dinosaur family tree."[144]

Premier vertebrate paleontologist, Alfred Romer also states,

> "It is obvious that it was forms of this sort from which arose the pterosaurs, birds and dinosaurs. There are no known thecodonts which show positive indications leading toward the first two groups mentioned, nor toward one of the two dinosaurian orders, the Ornithischia."[145]

Remember that the textbook mentioned above said that dinosaurs..."evolved from the crocodile-like reptiles known as thecodonts..." However, Romer, in his classic work, *Vertebrate Paleontology*, reports that..."There are no known thecodonts...." that can be shown through fossil evidence to qualify as ancestors of pterosaurs, birds or the entire order of Ornithischian dinosaurs.

The Horned Dinosaurs (Ceratopsians)

There are a number of different kinds of horned dinosaurs. Supposedly, from the evolutionary point of view, the protoceratopsids, which have no nose or brow horns, represent the earliest members of this particular group. Further, the fossil remains of *Leptoceratops* is their best candidate to occupy the position of the oldest, or most "primitive," of the protoceratopsids. However, as Lessem and Glut state, the features that characterize the fossils of *Leptoceratops* provide no substantial evidence for such claims,

> "Though it was one of the last of such dinosaurs, it was also very primitive, having a large head without a nasal horn. Since *Leptoceratops* is the most primitive known protoceratopsid, and yet one whose remains date from near the close of dinosaur time, it seems to be an ancestor that coexisted with its descendants."[146]

The fossils of *Leptoceratops* are found at the very "end" of the horned dinosaur era. However, since its fossils show it to possess very "primitive" features, it has to be considered as an ancestor of the horned dinosaurs. Since its fossils were found at the "end" of the horned dinosaur era, it was obviously a contemporary of the horned dinosaurs living in that era. It's so-called "primitive" features merely show it to have been another species of the ceratopsians.

Don Lessem, in his book, *Dinosaurs Rediscovered*, indicates that the fossils of the horned dinosaur, *Montanoceratops,* also contradict the assumed evolutionary progression of these "early" horned dinosaur forms,

> "Though it is supposedly the most derived of all proto-horned-dinosaurs, Weishampel found it in older rock than its less advanced cousin, *Leptoceratops*."[147]

Here again, are contradictions to the theory of evolution in the fossil record. *Montanoceratops* was "the most derived of all proto-horned-dinosaurs..." This means it was at the top of the evolutionary development of the proto-horned-dinosaur kind. Yet, Weishampel found it in sediments that were more ancient "according to the evolutionary dating scheme" than the "primitive" *Leptoceratops.*

The most popular of the horned dinosaurs is, of course, *Triceratops.* Probably more than 100 specimens have been excavated. It is supposed that these plant-eating dinosaurs evolved from the meat-eating dinosaurs. However, Dr. Angela Milner, of the Natural History Museum of London says,

"We are certainly lacking information that ties together meat-eating dinosaurs and all the rest of the dinosaurs. We certainly [don't know]. We really don't have any idea how the whole other group of dinosaurs, called ornithischians, [evolved from the meat-eating dinosaurs]. [We really do not know] exactly the timing and the way they branched off. We've got nothing there yet. There is a huge gap."[148]

Ornithopods

The ornithopoda are a large group of plant eating dinosaurs which include such species as the hypsilophodons, the iguanodons and the hadrosaurs. David Norman and Peter Wellnhofer, in their co-authored book, *The Illustrated Encyclopedia of Dinosaurs*, relate that,

"The earliest ornithopod (possibly a heterodontosaurid) is claimed to be an animal named *Pisanosaurus* from the late Triassic of Argentina."[149]

With regard to *Pisanosaurus*, there is significant division of opinion, as Catherine A. Forster, associate professor of biology at George Washington University, informs us that,

"However, some paleontologists are not convinced that Pisanosaurus is an ornithischian dinosaur, but feel that it, like the herrerasaurids, may have originated before the saurischian-ornithischian division."[150]

Evolution needs each animal group to have primitive ancestors. However, the fossil evidence is not supportive of this effort. With regard to *Pisanosaurus*, Dr. David Norman of Brasenose College, Oxford, reports in his book,

"However, the very fragmentary remains of this animal do not make it clear even if it is an ornithischian, let alone an ornithopod, so we should perhaps note the existence of this fossil and hope that more material is discovered that will prove exactly what sort of animal this is."[151]

Stegosaur

From the standpoint of evolution, with regard to *Huayangosaurus*, according to the book, *The Age of Dinosaurs* (consultant: Peter Dodson),

> "*Huayangosaurus* is the most primitive stegosaur."[152]

The fossils, however, throw confusing elements into the supposed evolutionary scenario. For instance,

> "At the front of its snout, *Huayangosaurus* had 14 teeth (seven on a side). Later stegosaurs lacked teeth in front."[153]

Wouldn't it seem more logical that these dinosaurs, in their theorized primitive state, would have begun without teeth in the front of their mouths and then, in the ensuing "millions of years" as they "progressed" to a higher and more sophisticated form, added more teeth so as to make their plant eating capacity more efficient? Would the loss of teeth, during the long "upward" evolutionary process, be considered a gain for a plant eating animal?

> "*Huayangosaurus* had long front limbs--almost as long as the back limbs. Later stegosaurs had forelimbs that were much shorter."[154]

Again, this may show variation within species, but there is no compelling argument that this shortening of their forelimbs over millions of years of time, provided any significant advantage for them.

Sauropods

After having acknowledged that the pterosaurs, birds, and Ornithischians have no thecodont ancestor, Romer then turns to deal with the other major group of dinosaurs, the Saurischians. This problem, he admits, is equally perplexing,

> "More troublesome is the situation with regard to the other dinosaur group, that of the Saurischia. As we shall see in the next chapter, the saurischians became abundant and varied before the end of the Triassic and, in their more primitive members, are not

far advanced structurally over the thecodonts. In consequence, it is difficult to establish a boundary between the two groups. Various Triassic forms, such *Ornithosuchus* and *Scleromochlus*, which were long classed as advanced thecodonts, are now regarded as having 'crossed the boundary' and are classed as primitive dinosaurs."[155]

The saurischians are divided into two sub groups: the sauropoda, which includes the long neck, long tail dinosaurs such as *Diplodocus* and the theropoda, which includes the flesh eating dinosaurs such as *Allosaurus*. John McIntosh is an acknowledged expert on sauropod dinosaurs. Below are a few of his statements which highlight the challenges to their supposed evolutionary origins,

"The origins of the Camarasauridae are obscure."[156]

"The subfamily Dicraeosauridae is somewhat of an enigma."[157]

"As here constituted, the Cetiosauridae represent a family of relatively generalized sauropods, all members of which are incompletely known at this time. That they comprise a monophyletic group is far from certain."[158]

Prosauropods

A group of dinosaurs, labeled prosauropods, have been put forward as being the ancestors of the sauropods. The fossil evidence for this is unsubstantial and has significantly undermined this particular evolutionary concept.

Regarding prosauropods, British paleontologist, Michael Benton, says,

"They were once thought to be ancestral to the giant sauropods, as their name indicates, but the idea is not quite correct."[159]

Vulcanodon is generally listed as a sauropod, however, the original description by Raath identified it as a prosauropod. The quotes below illustrate what a messy picture this has become. Michael R. Cooper comments,

"Since Raath's (1972) original description of *Vulcanodon Karibaensis* as a specialized melanorosaurid prosauropod, convergent towards the sauropod condition, a number of workers (Cruckshank 1975; Van Heerden 1978; Cooper 1980a) have discussed this species and concluded that it was in fact, a primitive sauropod."[160]

John McIntosh speaks with regard to this problem,

"This family diagnosis must be regarded as tentative since it incorporates some characters possessed by only one of the two genera, *Vulcanodon* and *Barapasaurus*, assigned to the family; the condition in the other is unknown."[161]

Lessem and Glut comment with regard to the dinosaurs *Barapasaurus*, the Titanosauridae, and the Melanorosauridae,

"*Barapasaurus* was a large browser with spoon-shaped teeth and relatively slender limbs...but how *Barapasaurs* looked and what it was remains uncertain. No skulls or feet have been found, and the proportions of the limbs are uncertain."[162]

"The least understood family of sauropods is the Upper Cretaceous Titanosauridae."[163]

"In 1965, Charig, Attridge, and Crompton broke with accepted dogma, that the sauropods had developed from bipedal prosauropod ancestors, and suggested that the ancestors were quadrupedal -- perhaps members of the little known Melanorosauridae."[164]

With regard to *Melanorosaurus*, Glut states that it

"...may actually be a sauropod rather than a prosauropod."[165]

David Norman suggests,

"The melanorosaurids may have represented an early attempt by prosauropods to use large size to become relatively invulnerable to predatory attack."[166]

Michael Benton also suggests,

"*Vulcanodon* could be related to *Melanorosaurus* or to *Cetiosaurus*."[167]

Dr. David Norman admits that,

"...the fossil record of prosauropods suffers many of the problems previously encountered in the theropods. Numerous species have been established upon totally inadequate fossil material."[168]

With regard to sauropod ancestry, J.S. McIntosh is unwilling to provide a cladogram for the sauropods,

"No cladogram of the sauropods is given, because I am convinced that one drawn ten years from now, when all the recently discovered material has been described and evaluated, would bear little resemblance to a current effort."[169]

"The picture then becomes confusing because of four factors splitting the sauropods into two classes. First, there is the division into the broad- and narrow- toothed forms, the latter compromising the Diplodocidae and apparently also the Titanosauridae. Second, there are the divided versus the undivided neural spines, the former compromising the Diplodocidae and Camarasauridae. Third, there are the normal versus the forked middle chevrons, the latter compromising the Diplodocidae and the three Chinese cetiosaurids. Fourth, there are those forms with 25-26 presacrals versus Chinese forms with 31-32 presacrals (*Euhelopus, Mamenchisaurus,* and *Omeisaurus*, each assigned to separate families). No cladogram can accommodate all of this data without admitting parallel development of some characters."[170]

Dinosaurs such as the plateosaurs have, in the past, been assigned the name "prosauropods" -- meaning that they were the theorized ancestors of the sauropods. However, Catherine A. Forster has stated,

"Exactly what the ancestors of prosauropods were, what they looked like, and where the prosauropods evolved is still a mystery.

> Although the name prosauropod, meaning "before-sauropods", implies they were the ancestors of the enormous sauropods, paleontologists now believe they did not give rise to the sauropods. They were already too specialized to have developed into the sauropods."[171]

Anthony R. Fiorillo, Ph.D., who is on staff with the Carnegie Museum of Natural History in Pittsburgh, Pennsylvania, is of the opinion that the prosauropods were the ancestors of the sauropods. Nevertheless, he acknowledges that,

> "One of the problems preventing the full acceptance of this ancestry, however, is in a comparison of the hind feet of prosauropods and sauropods. All prosauropods have a reduced fifth toe, a feature which is not reduced in the later sauropods. This may not seem, at first, to be significant. However, in the evolutionary development of most animals, once an anatomical feature is lost or reduced during the course of the history of that group of animals, the feature is not redeveloped at some later point. Additionally, there are aspects of the sauropod ankle which suggest to some workers that sauropods evolved independently of the prosauropods."[172]

According to Catherine A. Forster, the ancestors of the so-called prosauropods are also unknown,

> "The prosauropods and sauropods instead shared a common, yet unknown, ancestor, giving them a first cousin relationship."[173]

Further, paleontologist Jacques Van Heerden, asserts that the known prosauropods can not be the ancestors of the Sauropods,

> "Cruickshank (1975) and Van Heerden (1978) pointed out that the known prosauropods were already too specialized to have given rise to the Sauropoda and the latter author suggested that Palaeopoda be used instead..."[174]

In an interview with Dr. Carl Werner, Dr. David Weishampel of Johns Hopkins University stated,

> "Prosauropods actually are not on the direct line to get to sauropods, like *Diplodocus* and *Apatosaurus* and *Brachiosaurus*....They have their own evolutionary history that is independent of the evolutionary history of sauropods."[175]

Therapods

Finally, we have the group known as the therapods. The therapods are the carnivorous, or meat-eating, dinosaurs. These include, as mentioned above, all of the flesh eating dinosaurs such as *Allosaurus,* and *Tyrannosaurus.* So far, about 32 tyrannosaur specimens have been unearthed; however, no direct ancestors of this dinosaur have been discovered in the theorized time zones which preceded the *Tyrannosaurs.*

It has been theorized for a long time that these dinosaurs evolved from thecodont reptiles. The best two candidates which have been put forth as the "earliest" dinosaurs, those which are closest to the thecodont reptiles, are *Herrerasaurus* and *Eoraptor.* In the *Age of the Dinosaurs* (Consultant Peter Dodson),

> "*Herrarasuarus* represents the roots of dinosaur evolution. It was a theropod dinosaur, a distant ancestor of *Tyrannosaurus.* This early and very primitive dinosaur had four toes on its back feet. This separates it from other meat-eating dinosaurs, which had three toes."[176]

Don Lessem asserts that,

> "From fragmentary remains it was evident *Herrerasaurus* was four-toed and unusually short in the thigh for a dinosaur, both apparently primitive features."[177]

However, in both cases, these "earliest dinosaurs" show specializations which really do not qualify them as "missing links" or intermediates. They are, in a word, fully dinosaurian. For instance, with reference to *Herrerasaurus*, we find that,

"It was advanced in other ways, however. Its pelvis and back bones were similar to the advanced theropods of the Jurassic and Cretaceous Periods."[178]

David Lambert, popular writer of dinosaur books, states,

"In 1992 Paul Sereno and Fernando Novas proved *Herrerasaurus* was an early theropod, not a pre-saurischian, pre-ornithician dinosaur. The first dinosaurs must have lived even earlier, at least 230 million years ago."[179]

Further, in 1992, Catherine Forster comments with regard to the Herrerasaurid identity,

"Because they are not saurischians or ornithischians, some paleontologists originally argued (and still believe) that the herrerasaurids are not dinosaurs at all. Most paleontologists now feel that we simply need to stop considering the Dinosauria as being composed of only the Saurischia and Ornithischia. Since the herrerasaurids do have many "dinosaurian" features, they are considered to be the earliest, most primitive known dinosaurs. This early confusion as to whether herrerasaurids are dinosaurs or not shows how little we really know about the early evolution of these creatures, and in particular, the evolution of the dinosaurs before the saurischian-ornithischian split."[180]

As stated, *Coelophysis, Herrerasaurus* and *Eoraptor* are considered to be the ancestors of *T-Rex*. However, each of these have the wrong number of bones or the wrong shape of bones, which disqualifies them. Paul Sereno, paleontologist and professor at the University of Chicago, points out the problem of tyrannosaurid ancestry,

"Wherever we try to put *Tyrannosaurs* in the phylogeny of the branching history of the therapod dinosaurs, they have a long missing record. And we are going to find that record one of these days."[181]

The dating of the *Herrerasaurus* rocks has been based upon the subjective premise of dating the rocks by the fossils. Lessem, in his book, *Dinosaurs Rediscovered*, acknowledges that,

"The age of the Ischigualasto rock, only loosely estimated from the various animals fossilized within, indicated that *Herrerasaurus*, whatever it was, was among the earliest dinosaurs known."[182]

However, due to the fact that *Herrerasaurus* possessed some very unusual anatomical characteristics, such as in the shape of its neural spines,

"Such peculiarities put *Herrerasaurus* slightly off the main track of early dinosaur evolution, an early dinosaur that appeared, in some respects, to have gone its own peculiar evolutionary way."[183]

Consequently, Lessem further concludes,

"Just where *Herrerasaurus* fits into early dinosaur evolution is as yet uncertain."[184]

The so called "primitive" features which they possess are only assumed to be primitive. Since no one was there to see them, according to evolutionary theory, no one really knows that these were primitive features at all. According to evolutionary time reckoning, the earliest of the dinosaur sediments are the Triassic sediments. Therefore, it is theorized that the earliest dinosaurs must be found in those rocks. Those found in the lowest layers of these rocks are arbitrarily judged to have been the first dinosaurs to evolve. Any features that they possess which differ from the other dinosaurs are therefore judged to have been "primitive" features.

Do we, in fact, have a true blue ancestor of the therapods or just a few specimens of small fossilized dinosaurs, which is theorized to have filled that space? As Monastersky states in an article in *Science News*,

"On the basis of *Herrerasaurus* and other finds, researchers have pictured the earliest dinosaurs as small, bipedal carnivores -- a description that matches *Eoraptor*. This dinosaur itself could not have been the common ancestor because it had already evolved specializations that place it within the therapod subgroup. But the *Eoraptor* skeleton is as close to a common ancestor as researchers may ever find, Sereno says."[185]

Don Lessem reports that the "missing links" between the predinosaur and *Herrerasaurus* (a true dinosaur), are still missing,

"And somewhere between these predinosaur archosaurs and *Herrerasaurus*, these anatomical features evolved."[186]

To add to all of this uncertainty we have strong conviction on the part of some, such as Catherine Forster, that the herrerasaurids are neither saurischians nor ornithishians, and this creates a new problem,

"Because they are not saurischians or ornithischians, some paleontologists originally argued (and still believe) that the herrerasaurids are not dinosaurs at all. Most paleontologists now feel that we simply need to stop considering the Dinosauria as being composed of only the Saurischia and Ornithischia. Since the herrerasaurids do have many "dinosaurian" features, they are considered to be the earliest, most primitive known dinosaurs. This early confusion as to whether herrerasaurids are dinosaurs or not shows how little we really know about the early evolution of these creatures, and in particular, the evolution of the dinosaurs before the saurischian-ornithishian split."[187]

When accounts of any species' origins are written about by evolutionists they very often use the term that such and such a species "appears abruptly in the fossil record." What this means is that the fossil record provides no fossilized creatures which clearly support the idea that the organisms being discussed can be shown to have a series of transitional types which lead from the more simple species to the more complex. As the reader can see, such terminology is definitely applicable to the origin of dinosaurs. As Lessem and Glut remark in the *Dinosaur Encyclopedia* concerning the origin of the therapods, we read,

"Like sauropods and ornithischians, the earliest theropods appear suddenly in the fossil record. There is little hint of their origins except for their obvious but general anatomical resemblances to the earlier archosaurs...."[188]

Dinosaur Summary

Perhaps, as listed below, Dr. Weishampel's and Dr. Sereno's comments form a good summary of where we stand with regard to dinosaur ancestors,

> "From my reading of the fossil record of dinosaurs, no direct ancestors have been discovered for any dinosaur species. Alas, my list of dinosaurian ancestors is an empty one."[189]

> "Early on, again, I think researches and even maybe lay people really felt that we had more ancestors in the fossil record than we actually do...we don't have a lot of ancestors, we have a lot of twigs."[190]

In an article in *Nature*, evolutionist Barry Cox remarks,

> "Although many pages have been written discussing the mystery of the extinction of the dinosaurs, almost as much uncertainty surrounds their origin or origins."[191]

Speaking of dinosaurs in particular, and the process of linking fossils to time, Henry Gee, senior editor of *Nature*, comments,

> "Dinosaurs are fossils, and, like all fossils, they are isolated tableaux illuminating the measureless corridor of Deep Time. To recall what I said in chapter 1, no fossil is buried with its birth certificate. That, and the scarcity of fossils, means that it is effectively impossible to link fossils into chains of cause and effect in any valid way, whether we are talking about the extinction of the dinosaurs, or chains of ancestry and descent. Everything we think we know about the causal relations of events in Deep Time has been invented by us, after the fact."[192]

Science Versus

Though it has been believed that the ichthyosaur evolved from some terrestrial reptile, Romer insists that these creatures had no adaptations for land habitation,

> "These creatures were extreme in their marine adaptations, and their limbs were obviously unfitted for use on land."[193]

Author and science consultant, Steve Parker, admits,

> "The early evolution of the ichthyosaurs is not very clear."[194]

Romer, referring to the Ichthyosaurs and such, states,

> "We have no positive clues about the ancestry of the aquatic euryapsids as a whole."[195]

The late Professor Steven Jay Gould of Harvard makes an amazing statement with regard to the ichthyosaur,

> "This sea-going reptile with terrestrial ancestors converged so strongly on fishes that it actually evolved a dorsal fin and tail in just the right place and with just the right hydrological design. The evolution of these forms was all the more remarkable because they evolved from nothing -- the ancestral terrestrial reptile had no hump on its back or blade on its tail to act as a precursor."[196]

Gould's comment that the ichthyosaur "evolved from nothing" is not very complimentary to the dogmatism that characterizes evolutionary theory in general.

Now, concerning the origin of the Plesiosauria (long-necked plesiosaurs and short-necked pliosaurs). A report from the Bristol City Museum and Art Gallery informs us that,

"Although the fossil record tells us when the Plesiosauria first appeared and when they became extinct, it does not clearly show from which earlier reptile group they evolved."[197]

While opinion is divided on this subject, there are some who attempt to identify the nothosaurs as the ancestors of the pliosauria. However, Romer declares that even the ancestry of the nothosaurs is a mystery,

"The exact ancestry of the nothosaurs is uncertain."[198]

Mammal-Like Reptiles

Textbook Evolution

"By the end of the Permian Period, about 245 million years ago, a great variety of reptiles roamed Earth. One early group was the mammal-like reptiles, which displayed a mix of reptilian and mammalian characteristics. These chordates eventually came to dominate many land habitats. However, the mammal-like reptiles went extinct in just a few million years."[199] (Kenneth R. Miller, Joseph Levine, *Biology*, Prentice Hall)

"Mammals also first appeared during the late Triassic Period, probably evolving from mammallike reptiles. Mammals of the Triassic were very small, about the size of a mouse or shrew."[200] (*Ibid.*, p. 431)

Science Versus

Tom Kemp reports in the *New Scientist*,

"Each species of mammal-like reptile that has been found appears suddenly in the fossil record and is not preceded by the species that is directly ancestral to it. It disappears sometime later, equally abruptly, without leaving a directly descended species..."[201]

Regardless of their modern group name, mammal-like reptiles, Michael Denton asserts that the fossil evidence indicates,

> "The possibility that the mammal-like reptiles were completely reptilian in terms of their anatomy and physiology cannot be excluded. The only evidence we have regarding their soft biology is their cranial endocasts and these suggest that, as far as their central nervous systems were concerned, they were entirely reptilian."[202]

Reptile to Mammal

The constant assumption on the part of all these textbooks is that none of the basic animal kinds were created by a great Creator. Instead, they confidently assert that they evolve by totally naturalistic processes. This is the only perspective allowed.

Textbook Evolution

> "Mammals that had evolved during the Tertiary Period eventually faced a changing environment during the Quaternary Period."[203] (Kenneth R. Miller & Joseph Levine, *Biology*, Prentice Hall)

According to these textbooks, there is fossil evidence documenting the evolution of mammals,

> "According to the fossil record, the first ancestors of mammals diverged from ancient reptiles during the Permian Period, from 290 to 250 million years ago."[204] (*Ibid.*, p. 821)

Supposedly, the reptiles are the ancestors of mammals.

> "Therapsids, reptiles with complex teeth and legs positioned beneath their body, gave rise to mammals about the same time dinosaurs evolved, during the Triassic period."[205] (George B. Johnson, Peter H. Raven, *Biology*, Holt, Rinehart and Winston)

"Mammals are descended from ancient reptiles."[206] (Kenneth R. Miller & Joseph Levine, *Biology*, Pearson/Prentice Hall)

"Mammals probably evolved from warm-blooded reptiles."[207] (*California Life Science*, Pearson Education, Inc.)

Science Versus

According to Roger Lewin, the news editor of *Science* magazine for ten years,

"The (Evolutionary) transition to the first mammal, which probably happened in just one or, at most, two lineages is still an enigma."[208]

The evolutionary history of all mammals is unknown. Gordon Taylor, author of *The Great Evolution Mystery*, writes,

"...there are some twenty-six groups of mammals, the origins of which are completely obscure."[209]

Rodents make up approximately 40 percent of all mammals. It would seem that this would be the evolutionist's best category for finding at least some transitional forms. However, this is not the case. To begin with, Benton lets us know that,

"The phylogeny [evolutionary history] of rodents is controversial."[210]

Romer informs us that the evolutionary history of rodents is not merely controversial, but devoid of fossil ancestors,

"The origin of the rodents is obscure. When they first appear, in the late Paleocene, in the genus Paramys, we are already dealing with a typical, if rather primitive, true rodent, with the definitive ordinal characters well developed. Presumably, of course, they had arisen from some basal, insectivorous, placental stock; but no transitional forms are known."[211]

"There are a few fossil forms, back to the Myocene and possibly late Oligocene, but these give no indication of relationship of hystricids to other rodent types."[212]

Neither are there any transitional forms between the basic rodent types to be found in the fossil record,

"...the beavers are presumably derived from some primitive sciuromorph stock, but there are no annectant types between such forms and the oldest Oligocene castoroids to prove direct relationship."[213]

He goes on to say,

"Almost nothing is known of the ancestry or possible descendants."[214]

Regarding the origin of hares and rabbits, Romer states,

"The lagamorphs show no close approach to other placental groups, and the ordinal characters are well developed in even the oldest known forms."[215]

Horses

Textbook Evolution

"Changes in the structure of horse, for example, have been traced from the first appearance of a horselike mammal through various stages to what is now known as a horse."[216] (William D. Schraer & Herbert J. Stoltze, *Biology The Study of Life,* Prentice Hall, Fifth Edition)

"The fossil record provides information on the changes in the structure of the horse over time. During its evolution the horse has increased in size and the number of toes has decreased from four to just one."[217] (*Ibid.,* p. 580)

"Equus, the modern horse, evolved from the dog-sized hyracotherium. Notice the gradual transition, from Hyracotherium's four-toed front foot to the one-toed front foot of the modern horse."[218] (George B Johnson & Peter H. Raven, *Biology: Principles and Explorations,* Holt, Rinehart & Winston)

"...Modern horses are descendants of dog-sized animals with four toes on each front foot and three toes on each back foot -- modern horses have only one toe per foot. As you can see in figure 9.5, fossil intermediates between this 60 million-year-old ancestor horse and the modern horse reveal a history of slow transformation."[219] (George B. Johnson, *Biology Visualizing Life*, Holt, Rinehart & Winston)

These supposedly represent the fossil intermediates.

"*Figure 10.11* shows fossils of different horselike species. The oldest fossils (far left) are the leg and tooth of the small animal. How do the older fossils compare with the more recent fossils? This fossil record provides information about the evolutionary history of today's horse. If you study only the fossils below, there appears to be a straight-line progression from the first horse ancestor to the modern horse species. Such a progression implies an evolutionary "goal," since there is a trend toward larger body size, fewer toes, and flat teeth."[220] (Eric Strauss & Marylin Lisowski, *Biology: The Web of Life*, second edition, Scott Foresman-Addison Wesley)

Science Versus

For many years, textbooks and museums have shown pictures of the theorized horse series as if it was well documented by fossil evidence. A closer look at the literature and the fossils shows that this is not true at all. Four lines of evidence can be demonstrated against the theorized horse evolution series.

First, Hyracotherium is not a horse.

The first weak link in the horse evolution series is to be found in the mistaken assumption that the fossil animal called Eohippus is a horse.

Hyracotherium was discovered in Europe by England's leading anatomist, Dr. Richard Owen. It was assigned by him the genus name, Hyracotherium, because it most resembled the genus hyrax, which is not a horse, but a Coney or a rock badger. Later, other fossils of this same creature were discovered in North America. These were given the genus name, Eohippus by O.C. Marsh, Professor of Paleontology at Yale. Thus, we have two names for the same fossilized creature. However, the first name, Hyracotherium, therefore has priority. This means, in effect, that the name Eohippus should have been discarded. However, since it was necessary to establish an evolutionary sequence from one species to another that included the emergence of the true horse, the name Eohippus or "Dawn horse" was much more desirable to paleontologists oriented to the theory of evolution. So, we find that as a result of this creature being re-named Eohippus by a prestigious member of the American Paleontological Society, and because of a persuasive speech in favor of this scenario given by Thomas Huxley, a strong supporter of Darwin, the name Eohippus has been popularly accepted by evolutionists. This is so, even though Hyracotherium/Eohippus is not like a horse morphologically or by habitat.

One of the first to acknowledge this was G.A. Kerkut, in his *Implications of Evolution* (1960),

> "In the first place it is not clear that Hyracotherium was the ancestral horse. Thus Simpson (1945) states, 'Matthew has shown and insisted that *Hyracotherium* (including Eohippus) is so primitive that it is not much more definitely equid than tapirid, rhinocerotid, etc., but it is customary to place it at the root of the equid group.'"[221]

When closely examined, the fossils of Hyracotherium show that it had features that are exactly like those of today's living hyrax, but totally different from those of today's living horses. For instance, the fossil Hyracotherium/Eohippus had three toes both in the front and back feet. Further, the fossil and the living hyrax have arched backs, whereas, the horse always has sway backs. In the hyrax the eyes are midway between the nose and the ears, whereas, the horse had its eyes much closer to the ears. Finally, the teeth of fossil and living horses are found in the front and in the back of the mouth with a distinct gap separating the two sections (diasthema). By contrast the living hyrax and the fossil Hyracotherium/

Eohippus have a continuous line of teeth from the front to the back of the mouth. It is no wonder then that Nilsson takes the following position,

In his research, Nilsson was convinced that *Hyracotherium* was more similar to the Hyrax than to the horse.[222]

Steven Stanley, of the University of Hawaii at Manoa acknowledges,

> "Those who in the past have contemplated the formation of the modern horse by gradual evolution, beginning with this early genus, must now contend with the fact that at least two species of *Hyracotherium* lasted for several million years without appreciable change."[223]

> "It is notable that the evidence of great stability for species of *Hyracotherium* is complemented, at the other end of equid phylogeny, by data showing that ten species of horses lived through most or all of Pleistocene time -- for at least the better part of 2 Myr."[224]

Second, there are no transitional forms. Note this quote from Boyce Rensberger, writing in the *Houston Chronicle*

> "The popularly told example of horse evolution, suggesting a gradual sequence of changes from four-toed fox-sized creatures living 50 million years ago to today's much larger one-toed horse, has long known to be wrong. Instead of gradual change, fossils of each intermediate species appear fully distinct, persist unchanged, and then become extinct. Transitional forms are unknown."[225]

J.B. Birdsell, author of *Human Evolution*, informs us that,

> "The evolution of the foot mechanisms proceeded by rapid and abrupt changes rather than gradual ones. The transition from the form of foot shown by miniature *Eohippus* [*Hyracotherium*] to larger consistently three-toed *Miohippus* was so abrupt that it even left no record in the deposits...their foot structure changed very rapidly to a three-toed sprung foot in which the pad disappeared and the two side toes became essentially functionless. Finally, in

> the Pliocene the line leading to the modern one-toed grazer went through a rapid loss of the two side toes on each foot."[226]

Remember what the textbook told the students with regard to the evolution of the horse from the four-toed to the one-toed?

> "Equus, the modern horse, evolved from the dog-sized hyracotherium. Notice the gradual transition, from Hyracotherium's four-toed front foot to the one-toed front foot of the modern horse."[227]

The textbook statement above represents the party line with regard to the evolution of the horse. Unfortunately, the students never learned from these textbooks that in northeastern Oregon the three-toed horse (Neohipparion) is found in the same formation as the one-toed horse (Pliohippus).

With regard to the evolutionary reduction of toes, American Paleontologist, George G. Simpson, states,

> "The most famous of all equid trends, 'gradual reduction of the side toes' is flatly fictitious. There was no such trend in any line of Equidae."[228]

Third, horse fossils are not found in sequence.

Fossil horses are never found locally in their theorized sequence. There is no place in the world where one can go and see the horses in their succession from older to younger rocks. The various horse fossils appear abruptly at various locations without ancestors below or descendants above. In order to see these "so-called series" one would have to begin in North America, travel to Europe then back to America.

Fourth, fossil horses can be found in the wrong paleontological order.

J.T. Gregory, writing in *University of California Publications in Geological Sciences,* reports that there are no fossils there which would represent transitional species between the three-toed and one toed-horse. He also reminds us,

> "Elsewhere we have fossils of more primitive horses found in geological formations dated to be younger than formations containing more advanced species."[229]

Let's review again the textbook statement and compare it with that of a field researcher regarding the same subject. The Scott Foresman-Addison Wesley text tells the students, with regard to the horse series,

> "This fossil record provides information about the evolutionary history of today's horse. If you study only the fossils below, there appears to be a straight-line progression from the first horse ancestor to the modern horse species."[230]

By contrast, an expert on fossil horses, Bruce MacFadden of the University of Florida, while being confident of horse evolution, tells us the following,

> "...over the years fossil horses have been cited as a prime example of orthogenesis ['straight-line evolution']...it can no longer be considered a valid theory...we find that once a notion becomes part of accepted scientific knowledge, it is very difficult to modify or reject it"[231]

David Raup of the Field Museum of Natural History in Chicago speaks of the problems in relation to the evolution of the horse,

> "The record of evolution is still surprisingly jerky and, ironically, we have even fewer examples of evolutionary transition than we had in Darwin's time...Some of the classic cases of Darwinian change in the fossil record, such as the evolution of the horse, have had to be discarded...."[232]

In light of the above statement by David Raup, and in light of the statement below by the late George G. Simpson, a past authoritative spokesman for the evolutionary establishment, why, one wonders, is the horse series still so prominently mentioned in the state-adopted high school textbook?

> "The uniform, continuous transformation of Hyracotherium into Equus, so dear to the hearts of generations of textbook writers, never happened in nature."[233]

Way back in the 1960's, Kerkut and others reported on the evidence from the fossils that seriously undermined the case for horse evolution. Yet, in spite of these reports such as shown below, the high school textbooks present horse evolution as if it was all, paleontologically, as clear as a crystal,

> "One thing concerning the evolution of the horse has become clear. The story of the evolution of the horse has become more and more complex as further material is collected, and instead of a simple family tree the branches of the tree have increased in size and complexity till the shape is now more like a bush than a tree. In some ways it looks as if the pattern of horse evolution might be even as chaotic as that proposed by Osborn for the evolution of the Proboscidea, where, 'in almost no instance is any known form considered to be a descendant from any other known form; every subordinate grouping is assumed to have sprung, quite separately and usually without any known intermediate stage, from hypothetical common ancestors in the Early Eocene or Late Cretaceous."[234]

Kerkut also said,

> "The evolution of the horse provides one of the keystones in the teaching of evolutionary doctrine, though the actual story depends on who is telling it and when the story of the evolution of the horse. ...In the first place, it is not clear that Hyracotherium was the ancestral horse."[235]

Regarding the evolution of mammals in general, George Gaylord Simpson states that,

> "...The earliest and most primitive known members of every order already have the basic ordinal characters, and in no case is an approximately continuous sequence from one order to another known. In most cases the break is so sharp and the gap so large that the origin of the order is speculative and much disputed."[236]

Textbook Evolution

> "Fossil evidence shows that ancestors of the present-day whales and dolphins once lived on land."[237] (___, *Science Explorer: Focus on Earth Science*, Prentice Hall)

> "Whales are thought to have evolved from an ancestral line of four-legged mammals...."[238] (George B. Johnson & Peter H. Raven, *Biology,* Holt, Rinehart and Winston)

> "....Mesonychids are one hypothesized link between modern whales and certain hoofed mammals."[239] (*Ibid.*)

> "*Ambulocetus natans* apparently walked on land like modern sea lions and swam by flexing its backbone and paddling with its hind limbs (as do modern otters)."[240] (*Ibid.*)

Pakicetus

Regarding *Pakicetus*, we are informed by Philip Gingerich of Michigan University that,

> "Not far from the Khyber Pass in the arid Himalayan foothills of Pakistan, University of Michigan paleontologist Phillip D. Gingerich found a skull and several teeth and came to the startling conclusion that they belonged to an ancient walking whale."[241]

As referenced above, Gingerich considers this fossil to represent a walking whale, and therefore a transitional creature halfway between a whale and *Mesonyx*. The fossil, which was given the name *Mesonyx,* is considered to be a wolf-like mammalian carnivore. Regarding the proposed scenario of a wolf-like creature evolving into a whale by means of a transitional creature, which was somewhat whale-like and could swim and yet also possessed legs, it is important to be aware that only the cranium has been found for *Pakicetus*. No leg bones have ever been found.

According to the text book, *Modern Biology*,

"Scientists think that whales evolved from land-dwelling mammals. One of these ancestors may have belonged to the genus *Pakicetus*, which lived about 50 million years ago."[242]

Philip Gingerich, et al. admits,

"We do not yet know anything about the post-cranial anatomy of early Eocene whales."[243]

Basilosaurus

Textbook Evolution

"Figure 14-11: Fossil evidence suggests that ancient whales evolved from ancestors with hind limbs. The photograph shows the fossilized hind flipper of an early whale species, *Basilosaurus*."[244] (Neil A. Campbell, Brad Williamson, & Robin J. Heyden, *Biology: Exploring Life*, Pearson)

Science Versus

Basilosaurus was a very long skinny creature with front flippers and very small hind legs. There are some who believe that the back legs enabled the creature to walk, and others who believe they were used for reproduction. *Basilosaurus* is regarded by many evolutionists as a missing link in whale evolution; however, Dr. Lawrence Barnes who is with the Natural History Museum in Los Angeles does not agree. In an interview with Dr. Carl Werner, he stated that,

"...*Basilosaurus* existed at a time when baleen-bearing mysticetes [baleen whales] are known to have existed, and echolocating odontocetes [toothed whales] are presumed to have existed."[245]

If whales evolved, they evolved over a long period of time over a series of intermediates. Various authors are in the process of claiming certain fossilized specimens to be those theorized intermediates. The information in quotes found below will show the weaknesses of these arguments. Actually, according to E.H. Colbert, whales appear in the fossil record quite abruptly,

> "Like the bats, the whales appear suddenly in early Tertiary times, fully adapted by profound modifications of the basic mammalian structure for a highly specialized mode of life. Indeed the whales are even more isolated with relation to other mammals than the bats; they stand quite alone."[246]

Concerning Sirenia (sea cows) and Cetacea (whales, dolphins), the fossil evidence for their evolution is very weak, as noted by vertebrate paleontologist, Alfred Romer,

> "We are ignorant of their terrestrial forebears and cannot be sure of their place of origin."[247]

E.J. Slijper, an expert on whales, also comments on the lack of transitional forms,

> "We do not possess a single fossil of the transitional forms between the aforementioned land animals [i.e., carnivores and ungulates] and the whales."[248]

Despite the optimism with regard to *Ambulocetus* as referenced in the textbook above, Hans Thewissen, a whale paleontologist, along with his colleagues, Hussain, and Arif report that,

> "Since the pelvic girdle is not preserved, there is no direct evidence in *Ambulocetus* for a connection between the hind limbs and the axial skeleton. This hinders interpretations of locomotion in this animal, since many of the muscles that support and move the hindlimb originate on the pelvis."[249]

Ordinarily, *Ambulocetus* is drawn by artists to look like a "walking whale." It has an alligator-like snout with two front legs, two back legs and a relatively short tail. It has no whale-like aspects to its anatomy. However, as in the case of Hyracotherium in the famous horse series

(Hyracotherium was a hyrax and not a horse) the believers in evolution need to have an ancestor of the whale. Since, over time, they have believed that a cow, a wolf, or a hyena-like creature wandered back into the sea and over a period of millions of years became a whale, they needed to find the fossils of some extinct creature which had arms and legs, which could represent an early transition into a whale. *Ambulocetus* was thus chosen. In fact it was arbitrarily chosen and forced into the whale evolution series just as Hyracotherium/Eohippus was forced into the horse series. Note this acknowledgement by whale expert, Annalisa Berta:

> "*Ambulocetus* is a whale by virtue of its inclusion in that lineage."[250]

The eyes of the modern whale are located on the sides of their heads, as are the eyes of the theorized whale ancestor (the Hyena-like.... and the cougar-like, *Sinonyx*). *Ambulocetus*, however, had eyes on the top of its head like the crocodile. This has led Dr. Gingerich to state,

> "*Ambulocetus has its eyes raised up on top of its head in a very strange way, and it is unusually large for an early whale...maybe it's not on the main line* [in whale evolution]."[251]

Paleontologist, Barbara Stahl, informs us that,

> "The serpentine form of the body and the peculiar shape of the cheek teeth make it plain that these archaeocetes could not possibly have been the ancestor of modern whales."[252]

Furthermore, with regard to the theorized whale ancestor, *Sinonyx*, Dr. Gingerich shares,

> "...the [cat-like *Sinonyx* and the hyena-like *Pachyaena*]...will have to be put on a side branch...I doubt that they have any special relationship to whales."[253]

The third supposed intermediate between the cow-like creature and the whale is *Rodhocetus*. It is pictured as having legs, flippers and a small tail similar to that of a whale. However, the fossils now in hand of *Rodhocetus* do not include a whale's tail (fluke). Upon further investigation, according to

Dr. Carl Werner, the discoverer of *Rodhocetus*, Dr. Phil Gingerich at University of Michigan said the following in an interview:

> "I speculated that it might have had a fluke...I now doubt that *Rodhocetus* would have had a fluked tail."[254]

Other Mammals

Romer's research also reveals that the fossil record shows no specimens to qualify as ancestors for,

> 1) Old World porcupines (Hystricidae), "There are a few fossil forms, back to the Miocene and possibly late Oligocene, but these give no indication of relationship of hystricids to other rodent types."[255]
>
> 2) The rock rat (Petromus), "Almost nothing is known of the ancestry of *Petromus...*"[256]
>
> 3) Hares and rabbits, "The lagomorphs show no close approach to other placental groups, and the ordinal characters are well developed in even the oldest known forms."[257]
>
> 4) Conies, proboscideans, and sirenians, "... conies, proboscideans, and sirenians were already distinct groups at the time when they first appear in the fossil record."[258]

Origin of Flying Animals

Flying Reptiles

Textbook Evolution

In one of the Holt texts they show a picture of a fossilized pterosaur. Under the picture the words appear,

> "Fossils, such as this 150 million-year old pterosaur, a flying reptile, are one line of evidence for the occurrence of evolution."[259] (Holt, Rinehart and Winston, *Biology: Visualizing Life*)

Science Versus

Why does the above quote say that this pterosaur represents "one line of evidence for the occurrence of evolution"? The reason is that special creation by God has been totally ruled out. Thus, the only way, according to evolutionary theory, that a flying reptile could exist is that it must have evolved by naturalistic processes from a non-flying reptile. So, the mere discovery of a fossilized pterosaur is in itself "evidence" for evolution.

There are, in the fossil record, a variety of fossilized land dwelling reptiles and many flying reptiles. However, with regard to evidence of intermediates, E.C. Olsen, a Stanford Professor, reported

> "There is absolutely no sign of intermediate stages."[260]

According to Dr. Philip J. Currie of North Carolina State at Chapel Hill, talking about the origin of pterosaurs as follows,

> "When pterosaurs appeared in the fossil record, they were already highly specialized flying animals. And during their 140-million-year reign, they changed very little. No "missing link".....exists, making pterosaur origins more open to speculation."[261]

Paleontologist Kenneth Carpenter, director of the Prehistoric Museum in Price, Utah, states,

> "Thus, pterosaurs evolved flight for more than one reason. What is not yet known is whether flight evolved from a tree-climbing ancestor leaping from branch to branch, or from a ground runner chasing after insects. What this ancestor is we do not yet know."[262]

In other words, it is well understood that the origin of pterosaurs is a mystery, as a variety of researchers admit, including popular magazine and coffee table books, as noted below in an article by Carl Zimmer,

"How pterosaurs evolved toward flight is unknown..."[263]

The dinosaurologist, Paul Sereno, also acknowledges,

"For use in understanding the evolution of vertebrate flight, the early record of pterosaurs and bats is disappointing. Their most primitive representatives are fully transformed as capable fliers."[264]

Bats

Though it is believed that bats have evolved from some non-flying animal, there are no intermediates, as E.C. Olsen discloses, that demonstrate this theory,

"The first evidence of flight in mammals is in fully developed bats of the Eocene Epic."[265]

Romer says that when bats appear anywhere they are very clearly bats,

"Bats appear full fledged in both hemispheres in the Middle Eocene...."[266]

It is commonly recognized, as acknowledged by Carl Zimmer, that the evolutionary ancestry of bats is a challenge to the theory of evolution,

"The genealogy of bats is a murky matter, though, and has inspired some pretty exotic ideas,"[267]

"...bats pose one of evolution's supreme puzzles,"[268]

Colbert, Morales and Minkoff, in their book on the evolution of the vertebrates, make it clear that,

"There are no known intermediate stages between bats and insectivores,"[269]

Insects

Textbook Evolution

"From fossils, we know that insects were the first animals to evolve wings."[270] (George B. Johnson, *Biology Visualizing Life*, Holt, Rinehart & Winston)

Science Versus

Dr. E.C. Olson comments,

"There is almost nothing to give any information of the history of the origin of flight in insects."[271]

Dr. Pierre P. Grasse of the Sorbonne in France has been one of France's leading zoologists. He has written approximately seventeen books regarding the world of animals. With regard to insects, he states,

"We are in the dark concerning the origin of insects."[272]

Birds

Textbook Evolution

Listed below are the comments of the most prominent high school Biology textbook publishers. They include Holt, Rinehart, & Winston, Addison-Wesley, Glencoe, and Prentice Hall. They are all discussing the evolution of birds from reptiles,

"The success of birds lies in the evolution of a structure unique in the animal world -- the feather. Derived from reptilian scales, feathers are the ideal adaptation for flight..."[273] (G.B. Johnson and P.H. Raven, *Biology: Principles & Exploration,* Holt, Rinehart, & Winston)

"*Archaeopteryx* represents an intermediate form between reptiles and birds."[274] (Bernice Essenfeld, Carol Gontang, and Randy Moore, *Biology*, Addison-Wesley)

"Birds, such as *Archaeopteryx*, evolved from small bipedal dinosaurs about 150 million years ago but were not common until the flying reptiles called pterosaurs became extinct along with the dinosaurs."[275] (G.B. Johnson and P.H. Raven, *Biology: Principles & Exploration*, Holt, Rinehart, & Winston)

"While many gaps interrupted the fossil record in Darwin's era, even then, scientists knew of the *archaeopteryx* fossil transition between reptiles and birds."[276] (Raven & Johnson, *Biology*, 5th edition, McGraw-Hill)

"Recent fossil discoveries support the idea that modern birds evolved from some of these dinosaurs toward the end of this period. For example in **Figure 14.8**, you see the fossil of *Archaeopteryx*, a small dinosaur discovered in Germany. The fossil reveals that *Archaeopteryx* had feathers, a birdlike feature."[277] (Alton Biggs, Chris Kapicka, Linda Lundgren, *Biology: The Dynamics of Life*, Glenco/McGraw-Hill)

"To many paleontologists a bird is a dinosaur with feathers. That definition may sound odd, but it makes sense."[278] (Kenneth R. Miller, Joseph Levine, *Biology*, Prentice Hall)

"In fact, *Archaeopteryx* would be classified as a dinosaur except for one important feature: It had well-developed feathers covering most of its body. Because of this, the crow-sized *Archaeopteryx* is considered an early bird. Unlike modern birds, however, this creature had teeth in its beak, a bony tail, and toes and claws on its wings. Thus, *Archaeopteryx* is a transitional species with a combination of dinosaur-like and bird-like characteristics."[279] (*Ibid.*, p. 807)

"Birds apparently evolved from feathered dinosaurs during or after the Jurassic period."[280] (George B. Johnson, Peter H. Raven, *Biology*, Holt, Rinehart & Winston)

"Paleontologists agree that birds evolved from extinct reptiles."[281] (Kenneth R. Miller & Joseph Levine, *Biology*, Pearson/Prentice Hall)

"Most paleontologists think that birds evolved directly from dinosaurs."[282] (*Ibid.*)

"Thus, *Archaeopteryx* can be seen as a transitional species with characteristics of both dinosaurs and birds."[283] (*Ibid.*)

"Paleontologists think that *Archaeopteryx* and modern birds descended from some kind of reptile, possibly a dinosaur."[284] (Elizabeth Coolidge-Stolz, et al., *Life Science*, Pearson Prentice Hall)

Science Versus

Dr. W.E. Swinton, an expert on birds, truly believes that birds evolved from reptiles, but he admits that there is no fossil evidence to support this idea,

> "The origin of birds is largely a matter of deduction. There is no fossil evidence of the stages through which the remarkable change from reptiles to bird was achieved."[285]

In the past, *Archaeopteryx* has been promoted as being a very primitive and poor flyer. Non-flying birds have feathers different from those of flying birds. Research has shown, as W. K. Gregory of the American Museum of Natural History reported, that Archaeopteryx feathers are those of flying birds,

> "But in *Archaeopteryx*, it is noted, the feathers differ in no way from the most perfectly developed feathers known to us."[286]

Archaeopteryx is dismissed as a viable link between reptiles and birds by Gould and Eldridge,

> "...we are not even authorized to consider the exceptional case of the *Archaeopteryx* as a true link....An animal displaying characteristics belonging to two different groups cannot be treated as a true link as long as the mechanisms of transition remain unknown."[287]

The modern aspect of *Archaeopteryx* is highlighted by Douglas Palmer, in an article in *New Scientist*,

> "Remarkably, the wing feathers of the famous 150-million-year-old Late Jurassic bird, *Archaeopteryx,* are already well advanced and comparable to those of modern birds with an aerodynamic asymmetry and structural fabric that can evolve only as an adaptation for flight."[288]

The teeth were those of birds, as well as the brain and the inner ear. Stewart Martin of the University of Kansas, comments with regard to the teeth,

> "Therapod dinosaurs, by comparison, have serrated teeth with straight roots and no constriction....*Archaeopteryx* has unserrated teeth with constricted bases and expanded roots like those of other Mesozoic birds."[289]

Further, the presence of teeth in *Archaeopteryx* does not necessarily qualify it as the ancestor of modern birds. Scientist Paul Moody comments,

> "However, the other extinct birds had teeth, and every other category of vertebrates contains some organisms with teeth, and some without (amphibians, reptiles, extinct birds, mammals, etc.)."[290]

According to an article in *ScienceNOW* (4 Aug 2004),

> "Researchers at the University of Texas, Austin have used a computerized x-ray technique to build a three dimensional picture of the inside of the skull and inner ear of a fossil *Archaeopteryx*. From this they deduced *Archaeopteryx* had a brain and inner ear very similar to living birds...well suited for flight."[291]

Archaeopteryx is not understood to be reptilian, but very clearly, as E.C. Olsen affirms, "It shows itself to be a bird."[292]

Peter Dodson, American paleontologist with the University of Pennsylvania informs us that,

"At the end of the three days of presentations, Charig orchestrated a concerted effort to summarize the ideas for which consensus exists. The general credo runs as follows: *Archaeopteryx* was a bird that could fly, but it was not necessarily the direct ancestor of modern birds...."[293]

Alan Feduccia, of North Carolina State at Chapel Hill, an expert on fossil birds, concludes that,

"A search for ancestors in the fossil record is not likely to prove fruitful."[294]

It has been erroneous to conclude that the claws on *Archaeopteryx* represent a "primitive" feature. An article in *Scientific American* refers to the view expressed by Strahl, the president and CEO of the Chicago Zoological Society,

"Strahl adds that some ornithologists call the hoatzin 'primitive' because of its archaeopteryx-like claws; but he prefers to think of it as 'highly specialized.' Swans, ibis, and many other birds, he notes have wing claws; they just never make use of them."[295]

Now, fossils of a bird considered to be much older than *Archaeopteryx* has been found. *Science News* reports,

"In western Colorado's Dry Mesa Quarry, Brigham Young University archaeologists have come upon the 140 million-year-old remains of what they are calling 'the oldest bird ever found.'...."[296]

Since the fossil birds found by Chattergee are regarded as being millions of years older than *Archaeopteryx*, J. Ostrom of Yale acknowledges,

"It is obvious that we must now look for the ancestors of flying birds in a period much older than that in which *Archaeopteryx* lived."[297]

An article in *Science* Magazine notes,

"The hypothesis that *Archaeopteryx* represents a direct link from reptiles to birds has been generally accepted. The existence of another bird--one that was an adept flyer and thus more advanced on the evolutionary scale--would present a challenge to that hypothesis."[298]

Alan Feduccia says,

"Feathers are features unique to birds and there are no known intermediate structures between reptilian scales and feathers." [299]

Feduccia and Harrison B. Tordoff, a professor of zoology, state,

"The shape and general proportions of the wing and wing feathers in *Archaeopteryx* are essentially like those of modern birds."[300]

The modernity of *Archaeopteryx* is confirmed by Feduccia,

"I conclude that *Archaeopteryx* was arboreal and volant, considerably advanced aerodynamically, and probably capable of flapping, powered flight to at least some degree. *Archaeopteryx* probably cannot tell us much about the early origins of feathers and flight in true protobirds because *Archaeopteryx* was, in the modern sense, a bird."[301]

University of Kansas paleontologist, Larry Martin, remarked in 1985 that,

"*Archaeopteryx* is not ancestral to any group of modern birds." He further said that it is "the earliest known member of a totally extinct group of birds."[302]

Alan Feduccia, in the 6 quotes below, states,

"Just as important, and a discourse of the first part of this book, is that a dinosaurian origin of birds is inextricably linked with the cursorial, or ground-up origin, of avian flight, which is a biophysical impossibility. How could a large, deep-bodied, obligately bipedal reptile, with a long heavy, balancing tail and greatly foreshortened forelimbs, fighting gravity all the way, give rise to avian flight? The answer, of course, is that it could not."[303]

"Often the procedure is more like a board game than a careful scientific examination of anatomy, and simple tallies of characters by computer are used to determine who is related to whom."[304]

"Often there is simply insufficient evidence to offer such bold hypotheses, and cladograms confer an aura of truth to what is in reality speculation."[305]

"Paleontological cladists claim that opponents of the theropod origin must produce a more suitable ancestor, but alas, we simply don't have sufficient evidence."[306]

"The idea of a theropod origin of birds originally became coupled with the idea, now largely disproven, of hot-bloodedness or endothermy, in dinosaurs, the idea being that hot-blooded dinosaurs became coated with feathers for insulation, sprouted wing feathers on the arms, and flew off into the sunset."[307]

"However, although the theory of hot-bloodedness has fallen from favor, the advocates of a theropod origin of birds still reconstruct earthbound dinosaurs with a feathered coat, defying logic, as the feathers of birds that have lost the ability to fly invariably degenerate and become hairlike in appearance."[308]

In order for the promoters of evolution to make the case that dinosaurs evolved into birds, they first tried to establish the idea that dinosaurs were actually warm-blooded. That theory remains devoid of any significant evidence. Nevertheless, the high-school biology texts assert dogmatically that dinosaurs evolved into birds. Note this quote from the Holt, Rinehart, and Winston text,

"Birds apparently evolved from feathered dinosaurs during or after the Jurassic period."[309]

Again, Alan Feduccia, by contrast, very much disagrees with this idea,

"The idea that creatures could, over time, transform their body shape from heavy, deep bodied land animals with short forelimbs and large balancing tails into slight flyers is 'biologically impossible.'"[310]

Douglas Futuyma of the University of North Carolina states,

"Yale University released...a landmark publication...'the impetus for the book was the fact that this dinosaurian origin of birds had become, overnight, the dogma in the field, and many of us from the ornithological side felt that was simply wrong...that birds descended from dinosaurs is utter nonsense."[311]

The dinosaur to bird theory is a popular theory devoid of substantive fossil evidence. According to Pat Shipman of the department of Anthropology at Penn State,

"The bipedal ancestor hypothesis, while favored strongly by logic, has little direct evidence from the fossil record to support or refute it."[312]

In an article with other authors written in 2005, regarding the so-called feathered dinosaurs, Feduccia concludes that,

"...the problem of avian origins is far from being resolved."[313]

In a word, what we have is merely a theory without any significant evidence which would insure real confidence.

Biologist Edward Stuart Russell admits,

"Nothing is known with certainty as to how birds arose from reptiles or from what reptilian stock."[314]

Much fuss has been made about the discovery of fossils, which were supposed to be examples of dinosaurs with feathers. The most famous *archaeoraptor* was shown to be a composite fossil, that is, a hoax. In fact, the discovery of feathered dinosaurs seems to be a new lucrative business for some enterprising individuals.

Now, to add the the problem, we have fabricated fossils in the market, as associate editor of *Discover*, Kathy A. Svitil reports,

> "*Archaeoraptor* is just the tip of the iceberg. There are scores of fake fossils out there, and they have cast a dark shadow over the whole field. When you go to these fossil shows, it's difficult to tell which ones are faked and which ones are not. I have heard there is a fake-fossil factory in northeast China, in Liaoning Province, near the deposits where many of these recent alleged feathered dinosaurs were found."[315]

Summary

Darwin recognized from the very beginning that the fossil record did not strongly support his theory,

> "Geology assuredly does not reveal any such finely graduated organic change, and this is perhaps the most obvious and serious objection which can be urged against the theory."[316]

So the only thing that Darwin could do then was to predict that eventually the search for fossils would, sooner or later, lead to the discovery of transitional fossils which would verify his theory. This, the high-school textbook, *Biology* (Holt, Rinehart, & Winston), relates to the students,

> "Darwin predicted that intermediate forms between the great groups of organisms would eventually be found."[317]

However, the search for transitional fossils has not given any support to Darwin's theory. As Niles Eldredge of Columbia University and the American Museum of Natural History reports,

> "One hundred and twenty years of paleontological research later, it has become abundantly clear that the fossil record will not confirm this part of Darwin's predictions. Nor is the problem a miserably

> poor record. The fossil record simply shows that this prediction was wrong."[318]

After the authors of the Holt 2004 text (referenced above) tell the students about Darwin's prediction, they then immediately tell the students,

> "Since Darwin's time, many of these intermediaries have been found. For example, fossil intermediaries have been found between fishes and amphibians, between reptiles and birds, and between reptiles and mammals, adding valuable evidence about the fossil history of the vertebrates."[319]

The Prentice Hall text, *Biology*, states,

> "Darwin argued that the fossil record provided evidence that living things have been evolving for millions of years."[320]

The Glencoe text asserts,

> "Although the fossil record is incomplete, it does provide evidence that evolution has indeed occurred."[321]

Modern Biology, the Holt text, reports,

> "If today's species have come from more ancient forms, then we should be able to find remains of those species that no longer exist. Scientists have found such evidence in the form of fossils."[322]

We have referenced above the blatant statements of the high school textbooks telling the students that the fossil evidence favored evolution. In contrast to this propaganda found in the high school text, note what Mark Ridley of Oxford University, a leader in evolutionary thought, concludes from his research,

> "In any case, no real evolutionist...uses the fossil record as evidence in favor of the theory of evolution as opposed to special creation..."[323]

Why not? Because the fossil record does not contain any intermediate types. Theoretically, the idea of evolution involves slow and gradual change over millions of years of time. Change, incessant change, is the heart of the evolutionary process. The fossil record should document this process, but it does not. So the fossil record then becomes a liability to the evolutionists and must be minimized as to its importance to this theory.

Stasis, not change, as reported by Tom Kemp in *New Scientist*, is what characterizes the visible fossil record,

> "As is now well known, most fossil species appear instantaneously in the fossil record, persist for some millions of years virtually unchanged, only to disappear abruptly..."[324]

The late Stephen Gould of Harvard, states that significant changes are not observed,

> "Every paleontologist knows that most species don't change. That's bothersome....brings terrible distress....They may get a little bigger or bumpier but they remain the same species and that's not due to imperfection and gaps but stasis. And yet this remarkable stasis has generally been ignored as no data. If they don't change, its not evolution so you don't talk about it."[325]

He further comments,

> "Our ability to classify both living and fossil species distinctly and using the same criteria fit splendidly with creationist tenets. But how, he asks, could a division of the organic world into discrete entities be justified by an evolutionary theory that proclaimed ceaseless change as the fundamental fact of nature?"[326]

Also, David Raup of the Field Museum of Natural History in Chicago acknowledges,

> "The record of evolution is still surprisingly jerky and, ironically, we have even fewer examples of evolutionary transition than we had in Darwin's time."[327]

Even though the evolutionists firmly believe that they once existed, the fossil record does not provide any transitional forms from a lower species to a higher species. David Kitts, of the University of Oklahoma, confesses,

> "Despite the bright promise that paleontology provides a means of 'seeing' evolution, it has presented some nasty difficulties for evolutionists, the most notorious of which is the presence of 'gaps' in the fossil record."[328]

> "Evolution requires intermediate forms between species and paleontology does not provide them..."[329]

Niles Eldredge of Columbia University comments,

> "No wonder paleontologists shied away from evolution for so long. It never seems to happen. Assiduous collecting up cliff faces yields zigzags, minor oscillations, and the very occasional slight accumulation of change -- over millions of years, at a rate too slow to account for all the prodigious change that has occurred in evolutionary history. When we do see the introduction of evolutionary novelty, it usually shows up with a bang, and often with no firm evidence that the fossils did not evolve elsewhere! Evolution cannot forever be going on somewhere else. Yet that's how the fossil record has struck many a forlorn paleontologist looking to learn something about evolution."[330]

That is to say, in the fossil record there is no clear evidence of such a process. As Dr. Woodruff of University of California at San Diego explains,

> "But fossil species remain unchanged throughout most of their history and the record fails to contain a single example of significant transition."[331]

The problem of transitions is here addressed by Steven Stanley, formerly a paleontologist at Johns Hopkins University,

> "There are too many places where the fossil record is complete enough that we ought to see transitions occurring. Even in these cases we see very few good examples of higher taxa evolving by gradual change. There may be a few examples here and there, but by and large we just don't see the steps."[332]

David Swift of the University of Hawaii comments,

> "...no phylum can be traced from a preceding one in the fossil record, in fact we cannot account for the origin of a single phylum: they all appear abruptly. This is also true of lower taxonomic groups such as classes and orders, and possibly lower still."[333]

Steven Stanley acknowledges that,

> "The known fossil record fails to document a single example of phyletic evolution accomplishing a major morphologic transition."[334]

The lack of transitional fossils between one major type of organism and another has to be explained away so as to keep the theory of evolution viable. The reader will note in almost any evolutionary text that he reads on this subject that the terminology used will always be, "the abrupt appearance." Basically, this is a reluctant acknowledgement that the organisms in view have no visible ancestors in the fossil record; that is, there are no half-way or one-third of the way organisms between fish and amphibian or amphibian and reptile etc. As D. Dwight Davis, author of *The Giant Panda: a morphological study of evolutionary mechanisms*, acknowledged way back in 1949,

> "The sudden emergence of major adaptive types as seen in the abrupt appearance in the fossil record of families and orders, continued to give trouble. A few paleontologists even today cling to the idea that these gaps will be closed by further collecting, but most regard the observed discontinuity as real and have sought an explanation."[335]

Then, in 1953 G.G. Simpson, who was definitely one of the most influential paleontologists of the 20th century, admitted the same thing,

> "In spite of these examples, it remains true, as every paleontologist knows, that most new species, genera, and families, and that nearly all categories about the level of families, appear in the record suddenly and are not led up to by known, gradual, completely continuous transitional sequences."[336]

Later, in 1975 Valentine and Campbell, writing in *American Scientist,* also focused on this problem,

> "The abrupt appearance of higher taxa in the fossil record has been a perennial puzzle. Not only do characteristic and distinctive remains of phyla appear suddenly, without known ancestors, but several classes of phylum, orders of a class, and so on, commonly appear at approximately the same time, without known intermediates"[337]

References to "the abrupt appearance" have shown up in the literature in about every decade and will continue to do so for there are no transitional fossils leading from one simpler organism to a higher one because no such transitional organisms ever existed. However, when one is committed to an entirely materialistic explanation for the origin of species, and when a transcendent creator is inadmissible to the classroom, they are forced to use "words" in the place of evidence.

What is the explanation that will be used to resolve this embarrassing problem for the evolutionists? There is only one way out that they can take. Evolution has become true by definition, supported by consensus and not by evidence. Evidence has essentially become unnecessary. Thus, their explanation will be that there was no sedimentation occurring during the eras when the theoretical transitional species were alive. Gould explains,

> "The extreme rarity of transitional forms in the fossil record persists as the trade secret of paleontology. The evolutionary trees that adorn our textbooks have data only at the tips and nodes of their branches: the rest is inference, however reasonable, not the evidence of fossils. Yet Darwin was so wedded to gradualism that he wagered his entire theory on a denial of this literal record: 'The geological record is extremely imperfect and this fact will to a large extent explain why we do not find interminable varieties, connecting together all the extinct and existing forms of life by the finest graduated steps. He, who rejects these views on the nature of the geologic records, will rightly reject my whole theory.'"[338]

The late Stephen Gould of Harvard, as referenced above, acknowledges that the fossil record does not support Darwin's theory of gradualism (showing a gradual change in the fossil record from one lower species to a higher one), and, as Gould also relates, Darwin was well aware of this. In fact, he states that Darwin was so convinced that these intermediate fossils really existed, that he had to deny the "literal record." The literal record, in Darwin's day and now, is that there are no transitional fossils leading from any lower species to any higher species. Gould also states,

> "All paleontologists know that the fossil record contains precious little in the way of intermediate forms; transitions between major groups are characteristically abrupt."[339]

Jeffrey Schwartz, professor of biological anthropology at the University of Pittsburgh, comments,

> "Instead of filling in the gaps in the fossil record with so-called missing links, most paleontologists found themselves facing a situation in which there were only gaps in the fossil record, with no evidence of transformational intermediates between documented fossil species."[340]

Michael Denton notes,

> "There is a total absence of partially inclusive or intermediate classes, and therefore none of the groups traditionally cited by evolutionary biologists as intermediate gives even the slightest hint of a supposedly transitional character."[341]

Steven Stanley observes,

> "Established species are evolving so slowly that major transitions between genera and higher taxa must be occurring within small, rapidly evolving populations that leave no legible fossil record."[342]

If we buy into the neodarwinian picture of evolution with its slow and gradual changes over millions of years of time, we find that there are no transitional fossil links documenting the process of evolution.

Now, we find that if we buy into the new punctuated equilibrium theory of evolution, there is still no fossil record documenting the theory of evolution here either.

In either case there is no fossil record documenting the theory of evolution. Obviously the need to document the theory of evolution by use of the fossil record is no longer relevant to the theory of evolution. Remember again, Mark Ridley's statement already mentioned above,

> "In any case, no real evolutionist...uses the fossil record as evidence in favor of the theory of evolution as opposed to special creation..."

Newsweek magazine visited a conference of top evolutionists in 1980 and, with regard to the fossil record, reported the following,

> "Evidence from fossils now points overwhelmingly away from the classical Darwinism which most Americans learned in high school."[343]

As Stephen Gould and Niles Eldredge have said, "Phyletic gradualism [gradual evolution]...was never 'seen' in the rocks."[344] In other words, the scientific observation of the rock record was antagonistic to the theory of evolution right from the very beginning, but many paleontologists, other researchers and textbook writers would not acknowledge it.

If the idea of phyletic evolution was not seen in the rocks, even during the time of Darwin's writing, then where did this idea come from? According to Gould and Eldredge, it did not come from science but from the cultural and political biases of the 19th century,

> "It [gradual evolution] expressed the cultural and political biases of 19th century liberalism."[345]

In truth, the theory of evolution is supported primarily by consensus and imagination. N. A. Takahata, writing in *Annual Review of Ecology and Systematics*, notes,

> "Even with DNA sequence data, we have no direct access to the processes of evolution, so objective reconstruction of the vanished past can be achieved only by creative imagination."[346]

The original theory was evolution from one species to a higher species over millions of years of time, slowly and gradually. However, the lack of fossil evidence has forced modern evolutionists to change their theory. Instead of saying the lack of transitional fossils supports creation, since the creation model predicts that God created all of the basic kinds at the beginning, they must, at all costs avoid this capitulation. So, the new idea (punctuated equilibrium), simply explains away the lack of transitional fossils by stating that, during this theoretical period of transition, no sedimentation was being laid down, which would preserve the theoretical transitional fossils. The transition happened too rapidly. As Stanley relates,

> "The record now reveals that species typically survive for a hundred thousand generations, or even a million or more, without evolving very much. We seem forced to conclude that most evolution takes place rapidly....The point here is that if the transition was typically rapid and the population small and localized, fossil evidence of the event would never be found."[347]

It would never be found because it never existed.

However, at least one Prentice Hall textbook acknowledges the obvious difficulty with this view,

> "The supporters of punctuated equilibrium argue that transitional forms between species are missing because evolution occurs rapidly over a relatively brief period on the geologic time scale. Although the fossil record seems to support this new theory, the mechanisms that could produce new species in such a short interval are unknown."[348]

It is wholly illogical that there could occur a virtual rash of favorable mutations over a very geologically short period of time. According to the new theory of Punctuated Equilibrium, invertebrates lived for millions of years of time, then in a short period of time a few of them evolved into fish. Then the fish lived for millions of years in stasis (no change), and then in a short period of time a few of them evolved into amphibians; and, so this process continued all the way up to reptiles, mammals, primates and humans. This is probably the only theory in science that has ever been based on the lack of evidence. It speaks volumes of the bankruptcy of the theory of evolution, a theory that now must cite the lack of transitional fossils as a "proof" that evolution happened rapidly, as opposed to slow and gradually.

Dr. Colin Patterson, late director of the British Museum of Natural History, comments on this change of theories,

> "Well, it seems to me that they have accepted that the fossil record doesn't give them the support they would value so they searched around to find another model and found one....When you haven't got the evidence, you make up a story that will fit the lack of evidence."[349]

The late Stephen Gould expresses the revolutionary transition that has taken place within the evolution camp. Evolution has always been regarded as gradual change over time or "dissent with modification." However in the new theory necessitated by the intransigence of the fossil record to support evolution Gould states,

> "Stasis has become interesting as a central prediction of our theory."[350]

After over a hundred years of propaganda in the high school textbooks regarding the importance of the fossil record to the theory of evolution, in the end, it is found that it really offers no support at all. Even though Steven Stanley would not agree with my summary, in his book, *The New Evolutionary Timetable*, he nevertheless acknowledges,

> "In fact, the fossil record does not convincingly document a single transition from one species to another."[351]

If one challenges evolution in the public classroom the pat (and safe) response has been, "Evolution simply means change. Don't you believe in change?" So, for all these years, the central characteristic of evolution has been that of change. However, disappointing research in the fossil record has caused promoters of this theory to assert that the central characteristic is stasis (no change).

Niles Eldredge, writing in 1978, said in a news interview that he is, in a sense, forced to look for a new mechanism of evolution because in all their searching, "no one has found any evidence of such in between creatures as gradual evolution required."[352]

Science is what you see, not what you imagine, and the cold, hard facts with the regard to the theory of evolution is that it is imagined but never seen. This has been admitted by a number of scientists, including Stephen Gould of Harvard,

> "I regard the failure to find a clear 'vector of progress' in life's history as the most puzzling fact of the fossil record....we have sought to impose a pattern that we hoped to find on a world that does not really display it." [353]

> "The absence of fossil evidence for intermediary stages between major transitions in organic design, indeed our inability, even in our imagination, to construct functional intermediates in many cases, has been a persistent and nagging problem for gradualistic accounts of evolution."[354]

Niles Eldredge freely admits that the theorized transitional types have not been found,

> "No one has found any such in-between creatures. This was long chalked up to 'gaps' in the fossil records, gaps that proponents of gradualism [gradual evolutionary change from species to species] confidently expected to fill in someday when rock strata of the proper antiquity were eventually located. But all the fossil evidence to date has failed to turn up any such missing links. There is a growing conviction among many scientists that these transitional forms never existed"[355]

The Genesis account of Creation states that God created the various kinds of living organisms to reproduce "after their kind." That is what we see in the world today as reality, and the fossil record does not provide any theorized evolutionary "in-between creatures" which might contradict the Genesis account.

Gould further states,

> "We fancy ourselves as the only true students of life's history, yet to preserve our favored account of evolution by natural selection we view

our data as so bad that we never see the very process we profess to study."[356]

A logical conclusion to this statement is that this "process" is not a true historic phenomenon. It is really an imaginary event maintained by a means of consensus in order to support a preferred world-view.

Chapter One End Notes

[1] C.O. Dunbar *Historical Geology,* 2nd ed., John Wiley & Sons, 1960, p. 47
[2] Pierre P. Grasse, *The Evolution of Living Organisms,* Academic Press, 1977, p. 4
[3] Steven M. Stanley, *Macroevolution: Pattern and Process,* W.H. Freeman, 1979, p. 2
[4] George B. Johnson and Peter H. Raven, *BIOLOGY:Principles and Explorations*, Holt, Rinehart and Winston, Inc. 1996, p.252
[5] Michael Padilla, Ioannis Miaoulis, Martha Cyr, *California Focus on Life Science*, Pearson/Prentice Hall, 2008, p. 236
[6] Sir Archibald Geikie, *Textbook of Geology*, Reprint 2010, p. 929
[7] Philipp H. Kuenan, *Marine Geology*, John Wiley & Sons, 1950, p. 548
[8] Alton Biggs, Chris Kapicka, Linda Lundgren, *Biology: The Dynamics of Life*, Glencoe, McMillian/ McGraw-Hill, 1995, p. 431
[9] George B. Johnson, *Biology:Visualizing Life*, Holt, Rinehart and Winston, 1994, p. 181
[10] Sylvia S. Mader, *Biology,* McGraw Hill, 2001, p.292
[11] George B. Johnson and Peter H. Raven, *BIOLOGY*, Holt, Rinehart and Winston, Inc. 2004, p.283
[12] Alton Biggs, et al., *Biology: The Dynamics of Life*, Glenco, 2004, p. 399-400
[13] K.R. Miller, J. Levine, *Biology,* Prentice Hall, 2006, p. 383
[14] John H. Postlethwait & Janet L. Hopson, *Modern Biology*, Holt, Rinehart, & Winston, 2006, p. 302
[15] Alton Biggs, et. al., *Biology: The Dynamics of Life*, Glencoe Science, *National Geographic*, McGraw-Hill/Glencoe, 2006, pp. 399-400
[16] Alton Biggs, Whitney Crispen Hagins, William G. Holliday, Chris L. Kapicka, Linda Lundgren, Ann Haley MacKenzie, William D. Rogers, Marion B. Sewer, Dinah Zike, *Biology,* Glencoe, 2007, p. 424
[17] George B. Johnson, *Biology Visualizing Life*, Holt, Rinehart & Winston, 1994, p. 56
[18] *Ibid.*
[19] *Ibid.*
[20] *Ibid.*
[21] Kenneth R. Miller and Joseph Levine, *Biology*, Prentice Hall, 2002, p. 429
[22] L. Oliwenstein, "Onward and Upward?", *Discover* magazine, June 1993, p. 22
[23] Steven M. Stanley, Macroevolution: Pattern and Process, W.H. Freeman, 1979, p. 69
[24] Preston Cloud, "Pseudofossils, A Plea for Caution" in *Geology*, November 1973, p. 123
[25] Andrew H. Knoll, "End of the Proterozoic Eon," *Scientific American*, vol. 265 [October 1991], p. 64-65
[26] George B. Johnson, *Biology Visualizing Life*, Holt, Rinehart & Winston, 1994, p. 56
[27] *Ibid.*
[28] Jacques Monod, *Chance and Necessity*, New York: Alfred A. Knoph, p.134
[29] Kenneth R. Miller and Joseph Levine, *Biology*, Prentice Hall, 2002, p. 417
[30] Sylvia S. Mader, *Biology,* McGraw Hill, 2001, p.4
[31] Anonymous, "Ancient Alga Fossil Most Complex Yet," *Science News*, vol. 108, [September 20, 1975], p. 181
[32] *Ibid.*
[33] S. M. Stanley, *Johns Hopkins Magazine*, p.6, June, 1982
[34] Richard Dawkins, *The Blind Watchmaker*, 1986, p.116
[35] Anonymous, "Ancient Alga Fossil Most Complex Yet," *Science News*, Vol. 108, September 20, 1975, p. 181
[36] *Ibid.*
[37] *Ibid.*
[38] Stefan Bengston, "The Solution to a Jigsaw Puzzle," *Nature*, vol. 345 [June 28, 1990], p. 765
[39] Alton Biggs, *et. al*, *Biology,* Glencoe, 2007, p. 424
[40] *Ibid.*
[41] Sylvia S. Mader, *Biology*, McGraw Hill, 2001, p.329
[42] Kenneth R. Miller & Joseph Levine, *Biology*, Pearson/Prentice Hall, 2004, p. 421

[43] Stephen J. Gould, *Nature*, Vol. 377, 26 10/95, p. 682
[44] Marshall Kay & Edwin H. Colbert, "Stratigraphy and Life History," John Wiley & Sons, 1965, p. 102
[45] Stephen Jay Gould, "A Short Way to Big Ends," *Natural History,* vol. 95 [January 1986], p. 18
[46] Stephen Jay Gould, *The Panda's Thumb*, 1980, pp. 238-239
[47] Darwin, Charles, *On the Origin of Species.* 1859, p. 308
[48] Thomas Neville George, "Fossils in Evolutionary Perspective," *Science Progress*, vol. 48, January 1960, p. 5
[49] Eugene Koonin, "The Biological Big Bang Model for the Major Transitions in Evolution," *Biology Direct*, 2007, 2:21.
[50] Richard Monastersky, "When Earth Tipped, Life Went Wild," *Science News*, vol. 152, 1997, p. 52
[51] Stephen J. Gould, "The Evolution of Life," in Schopf, *Evolution: Facts and Fallacies*, 1999, p. 9
[52] Richard Dawkins, *The Blind Watchmaker*, 1986, p. 229
[53] J. Madeleine Nash, "When Life Exploded", *TIME*, December 4, 1995, p.68
[54] Richard Dawkins, *The Blind Watchmaker,* New York: W. W. Norton, 1987, p. 229
[55] Lori Oliwenstein, 'Onward and Upward?', *Discover* magazine, June 1993, p. 22
[56] Derek E. G. Briggs, "Extraordinary Fossils," *American Scientist*, vol. 79 [March/April], 1991, p. 135
[57] *Ibid.,* p. 136
[58] *Ibid.,* p. 137
[59] David M. Martill, "Preservation of Fish in the Cretaceous Santana formation of Brazil," *Paleonatology*, Vol. 31, Part 1, 1988, p. 1
[60] Dale A. Russell, *An Odessey in Time: The Dinosaurs of North America*, NorthWord Press, Inc., 1989, p. 46
[61] Carl Zimmer, "Masters of an Ancient Sky," *Discover,* February, 1994, pp. 44-45
[62] Alton Biggs, Chris Kapicka, Linda Lundgren, *Biology*, *The Dynamics of Life*, Glencoe, McMillian/ McGraw-Hill,1995, p. 406
[63] Derek E.G. Briggs, "Extraordinary Fossils," *American Scientist*, vol. 79 [March/April], 1991, p. 139
[64] James W. Valentine and Douglas H. Erwin, "Interpreting Great Developmental Experiments: The Fossil Record", *Development as an Evolutionary Process,* Alan R. Lias Inc., 1987, p. 84
[65] James W. Valentine & Cathryn A. Campbell, "Genetic Regulation and the Fossil Record," in *American Scientist*, Vol. 63, November-December, 1975, p. 673
[66] Derek Ager, *Proc. Geol. Assoc.,* vol. 87, 1976, p. 132
[67] H. S. Ladd *Geology South of American Me*, 1967, Vol. 2, p. 7
[68] Hickman, Roberts, and Larson, *Zoology*, W. C. Brown, 1997, p. 275
[69] *Ibid.,* p. 275
[70] *Ibid.,* p. 399
[71] *Ibid.,* p. 450
[72] Richard Monastersky, Earth Science Ed., Science News, *Discover*, 4/93, p. 40
[73] Daniel I. Axelrod, UCLA, *Science*, Vol. 200, 5/1978, p. 529
[74] Richard Monastersky, Earth Science Ed., Science News, *Discover*, 4/93, p. 40
[75] Douglas Futuyma, *Evolutionary Biology*, 1985, p. 325
[76] Jeffrey S. Levinton, "The Big Bang of Animal Evolution," *Scientific American*, vol. 267, no. 5, November 1992, p. 52
[77] Bernice Essenfeld, Carol Gontang, Randy Moore, *Biology*, Addison-Wesley, 1994, p. 554
[78] George B. Johnson, *Biology Visualizing Life*, Holt, Rinehart & Winston, 1994, p. 211
[79] Bernice Essenfeld, Carol Gontang, Randy Moore, *Biology*, Addison-Wesley, 1994, p. 556
[80] Kenneth R. Miller & Joseph Levine, *Biology*, Pearson/Prentice Hall, 2008, p. 772
[81] Padilla, Michael, et al., *California Focus on Life Science*, Pearson Prentice Hall, 2008, p. 289
[82] Francis Hitching, *The Neck of the Giraffe*, Macmillan, 1982, p. 10
[83] F.D. Ommanney, *The Fishes* , Life Nature Library, 1964, p. 60
[84] A.S. Romer, *Man and the Vertebrates*, Penguin Books, 1968, p. 16
[85] Hickman, Roberts, and Larson, *Zoology,* W. C. Brown, 1997, pg. 480
[86] Mustaga Kuru, *Omurgali Hayvanlar* (Vertebrates), Gazi University Publications, 5th ed., Ankara, 1996, p. 27

[87] David M. Raup, "On the Early Origins of Major Biologic Groups," *Paleobiology*, vol. 9, no. 2, p. 107
[88] Alfred Sherwood Romer, "Major Steps in Vertebrate Evolution," *Science,* vol. 158 [December 29, 1967], p. 1632
[89] Bruce Runnegar and John Pojeta, Jr., "Molluscan Phylogeny: The Paleontological Viewpoint," *Science*, Vol. 186, October 25, 1974, p. 316
[90] James Hopson and Leonard B. Radinsky "Vertebrates Paleontology: New Approaches and New Insights," *Paleobiolgoy*, Vol. 6, Summer 1980, p. 258
[91] Robert L. Carroll, *Patterns and Processes of Vertebrate Evolution*, Cambridge University Press, 1997, p. 296
[92] Gerald T. Todd, "Evolution of the Lung and the Origin of Bony Fishes: A Causal Relationship?" *American Zoologist*, vol.20, no. 4, 1980, p. 757
[93] Alton Biggs, Chris Kapicka, Linda Lundgren, *Biology, The Dynamics of Life*, Glencoe, McMillian/ McGraw Hill, 1995, p. 770
[94] *Ibid.*, p. 771
[95] Colin Tudge, "Throw Away Your Zoology Textbooks," *New Scientists* [November 8, 1973], p. 393
[96] Barbara Stahl, 1974. *Vertebrate history: Problems in evolution*, New York: Dover Publications, Inc., p. 126
[97] E.H. Colbert, M. Morales, and E. C. Minkoff. 2001, *Evolution of the Vertebrates: A history of the backboned animals through time*, 5th ed., New York, Wiley-Liss, Inc., p. 53
[98] J.O. Long, *The Rise of Fishes: 500 Million Years of Evolution*, John Hopkins University Press, 1995, p. 30
[99] *Ibid.*, p. 66
[100] *Ibid.*, p. 69
[101] *Ibid.*, p. 91
[102] Errol White (F.R.S.), *Proceedings of the Linean Society of London,* Vol.177, p. 1, Jan. 1996
[103] C.P. Hickman, L.S. Roberts & A. Larson, *Integrated Principles of Zoology*, 2001, p. 151
[104] J.R. Norman, Department of Zoology, British Museum of Natural History, *A History of Fishes,* 1975, p. 343
[105] George B. Johnson and Peter H. Raven, *Biology: Principles and Explorations*, Holt, Rinehart and Winston, Inc., 1996, p. 703
[106] *Ibid.*
[107] *Ibid.*, p. 745
[108] *Ibid.*, p. 744
[109] George B. Johnson and Peter H. Raven, *Biology: Principles and Explorations*, Holt, Rinehart and Winston, Inc., 1996, p. 706
[110] Alton Biggs, Chris Kapicka, Linda Lundgren, *Biology*, *The Dynamics of Life*, Glencoe, McMillian/ McGraw-Hill, 1995, p. 817
[111] *Ibid.*, p. 830
[112] G.B. Johnson and P.H. Raven, *Biology: Principles & Exploration*, Hold, Rinehart and Winston, Inc., 1996, p 756
[113] Bernice Essenfeld, Carol Gontang, Randy Moore, *Biology*, Addison-Wesley, 1994, p. 573
[114] *Ibid.*
[115] G.B. Johnson and P.H. Raven, *Biology: Principles & Exploration*, 1996, p 732
[116] Hans Fricke, "Coelacanths: The Fish that Time Forgot" *National Geographic* Vol. 173, No. 6, June 1988, pp. 824-838
[117] *Modern Biology*, Holt, Rinehart, & Winston, 2002, p. 819
[118] Kenneth R. Miller and Joseph Levine, *Biology*, Pearson/Prentice Hall, 2008, p. 782
[119] Barbara J. Stahl, *Problems in Evolution*, 1974, p.148
[120] *Ibid.*
[121] Gordon R. Taylor, *The Great Evolution Mystery,* New York: Harper & Row, 1983, p. 60
[122] James A. Hopson and Leonard B. Radinsky, "Vertebrate Paleontology," *Paleobiology*, Vol. 6, No. 3, Summer 1980, p. 258
[123] Robert Carroll, *Vertebrate Paleontolgoy and Evolution*, New York: W.H. Freeman and Co., 1988, p. 138

[124] Pascale Chenel, *Life and Death of Dinosaurs*, Barron's, 1987 (first English edition), p. 37
[125] *National Geographic*, Vol. 173, No. 6, 1988, pg. 837
[126] Robert L. Carroll, *Vertebrate Paleontology and Evolution*, New York: W.H. Freeman and Co., 1988, p. 181-184
[127] Edwin H. Colbert, and M. Morales, *Evolution of theVertebrates*, New York: John Wiley and Sons, 1991, p. 99
[128] James, Hopson and Leonard B. Radinsky, "Vertebrates Paleontology: New Approaches and New Insights," *Paleobiology,* vol. 6, Summer 1980, p. 258
[129] Malcolm W. Browne, "Biologists Debate Man's Fishy Ancestors," *New York Times*, March 6, 1993
[130] Alton Biggs, Chris Kapicka, Linda Lundgren, *Biology*, *The Dynamics of Life*, Glencoe, McMillian/ McGraw-Hill, 1995, p. 844
[131] *Modern Biology,* Holt, Rinehart, & Winston, 2002, p. 839
[132] George B. Johnson, Peter H. Raven, *Biology*, Holt, Rinehart and Winston, 2004, p. 268
[133] Kenneth R. Miller & Joseph Levine, *Biology*, Pearson/Prentice Hall, 2008, p. 798
[134] Robert Carroll, "Problems of the Origin of Reptiles," *Biological Reviews of the Cambridge Philosophical Society*, Vol. 44, No. 3 July 1969, p. 393
[135] Edwin H. Colbert, and M. Morales, *Evolution of the Vertebrates*, John Wiley and Sons, 1991, p. 223
[136] C.O. Dunbar *Historical Geology,* 2nd ed. p. 47 John Wiley & Sons, New York, 1960
[137] M.J. Benton, 2005, *Vertebrate Palaeontology*, Malden, MA: Blackwell Science, p. 244
[138] E.H. Colbert and M. Morales, 1991, *Evolution of the Vertebrates*, John Wiley and Sons, p. 216
[139] George B. Johnson and Peter H. Raven, *Biology: Principles & Exploration*, Holt, Rinehart and Winston, 1996, p. 736
[140] *Modern Biology*, Holt, Rinehart, & Winston, 2002, p. 839
[141] George B. Johnson, Peter H. Raven, *Biology*, Holt, Rinehart and Winston, 2004, p. 268
[142] Alton Biggs, et. al., *Biology: The Dynamics of Life*, Glencoe Science, *National Geographic*, McGraw-Hill/Glencoe, 2006, p. 819
[143] Dale A. Russell, *The Dinosaurs of North America*, North Word Press, Inc. In association with National Museum of Natural Sciences, 1989 p. 60
[144] Don Lessem & Donald F. Glut, *Dinosaur Encyclopedia*, Random House, Inc.,1993, p. xi
[145] Alfred Sherwood Romer, *Vertebrate Paleontology*, 3rd ed., University of Chicago Press, 1966, p. 140
[146] Don Lessem & Donald F. Glut, *Dinosaur Encyclopedia*, Random House, 1993, p. 268-269
[147] Don Lessem, *Dinosaurs Rediscovered,* A Touchstone Book, Simon & Schuster, 1992, p. 285
[148] Interview with Dr. Angela Milner, Paleontologist and Head of Vertebrate Paleontology, Natural History Museum of London, for video series *Evolution: The Grand Experiment*, conducted November 16, 1998, by Dr. Carl Werner
[149] Dr. David Norman and Dr. Peter Wellnhofer, *The Illustrated Encyclopedia of Dinosaurs*, Salamander Books Limited, 2000, p. 98
[150] Catherine A. Forster, "The First Dinosaurs," found in *The Ultimate Dinosaur*, Edited by Byron Preiss and Robert Silverberg, Bantam Books, 1992, p. 31
[151] David Norman, *The Illustrated Encyclopedia of Dinosaurs*, Crescent Books, 1995, p. 98
[152] ______, Consultant: Peter Dodson, *The Age of Dinosaurs*, Publications International, 1993, p. 71
[153] *Ibid.*
[154] *Ibid.*
[155] Romer, *Vertebrate Palentology*, 3rd ed., University of Chicago Press, 1966, p. 140
[156] John McIntosh, *Dinosaur Systematics, Approaches and Perspectives,* edited by Kenneth Carpenter and Philip J. Currie, Cambridge University Press, 1992, p.55
[157] *Ibid.,* p. 56
[158] John S. McIntosh, "Sauropoda," *The Dinosauria*, Edited by David B. Weishampel, Peter Dodson, Halszka Osmolska, University of California Press, 1990, p. 377
[159] Michael Benton, *The Dinosaur Encyclopedia*, Simon Pulse, 1984, p. 101
[160] Michael R. Cooper, "A Reassessment of V*ulcanodon Karibaensis* Raath (Dinosauria:Saurischia) and the Origin of the Sauropoda," *Palaeont.* Afr., 25, 203-231, 1984

[161] J.S. McIntosh, "Sauropoda," *The Dinosauria*, 1990, p. 377
[162] Don Lessem and Donald F. Glut, *Dinosaur Encyclopedia*, Random House, 1993, p. 61
[163] *Ibid.*, p. 57
[164] *Ibid.*, p. 54
[165] Donald F. Glut, *The New Dinosaur Dictionary*, Citadel Pr, 1982, p. 172
[166] Dr. David Norman, *The Illustrated Encyclopedia of Dinosaurs*, Salamander Books, 2000, p. 79
[167] Michael Benton, "*Dinosaurs, An A-Z Guide,*" Random House Value Publishing, 1988, p. 164
[168] Dr. David Norman, *The Illustrated Encyclopedia of Dinosaurs*, Salamander Books, 2000, p. 74
[169] J.S. McIntosh, *Op Cit,* p. 57
[170] J.S. McIntosh, *Dinosaur Systematics: Approaches and Perspectives*, edited by Kenneth Carpenter & Philip J. Currie, 1992, 57
[171] Catherine A. Forster, "The First Dinosaurs," found in *The Ultimate Dinosaur*, Edited by Byron Preiss and Robert Silverberg, Bantam Books, 1992, p. 34
[172] Anthony R. Fiorillo, "The Age of Giants," found in *The Ultimate Dinosaur*, Edited by Byron Preiss and Robert Siverberg, Banton Books, 1992, p.106
[173] Catherine A. Forster, "The First Dinosaurs," found in *The Ultimate Dinosaur*, Edited by Byron Preiss and Robert Silverberg, Bantam Books, 1992, p. 34
[174] Jacques Van Heerden, "The Morphology and Taxonomy of *Euskelosaurus* (Reptilia: Saurischia; Late Triassic) From South Africa," Vol. 4, December 1979
[175] Dr. Carl Werner, *Evolution: The Grand Experiment*, New Leaf Press, 2007, p. 125
[176] ______, Consultant: Peter Dodson, *The Age of Dinosaurs*, Publications International, 1993, p. 36
[177] Don Lessem, *Dinosaurs Rediscovered*, A Touchstone Book, Simon and Schuster, 1992, p. 62
[178] ______, Consultant: Peter Dodson, *The Age of Dinosaurs*, Publications International, 1993, p. 36
[179] David Lambert, *The Ultimate Dinosaur Book*, Dorling Kindersley, 1993, p. 41
[180] Catherine A. Forster, "The First Dinosaurs," found in *The Ultimate Dinosaur*, Edited by Byron Press and Robert Silverberg, Bantam Books, 1992, p. 31
[181] Interview with Dr. Paul Sereno, for video series *Evolution: The Grand Experiment*, conducted February 24, 1999, by Dr. Carl Werner
[182] Don Lessem, *Dinosaurs Rediscovered*, A Touchstone Book, Simon and Schuster, 1992, p. 62
[183] *Ibid.*, p. 72
[184] *Ibid.*
[185] Richard Monastersky, "The Accidental Reign," *Science News*, Vol. 143, 1993, p. 61
[186] Don Lessem, *Dinosaurs Rediscovered*, A Touchstone Book, Simon and Schuster, 1992, p. 62
[187] Catherine A. Forster, "The First Dinosaurs," found in *The Ultimate Dinosaur*, Edited by Byron Press and Robert Silverberg, Bantam Books, 1992, p. 31
[188] Don Lessem & Donald F. Glut, *Dinosaur Encyclopedia*, Random House, Inc.,1993, p. xii
[189] Dr. David Weishampel as quoted by Dr. Carl Werner, *Evolution: The Grand Experiment*, New Leaf Press, 2007, p. 126
[190] Dr. Paul Sereno as quoted by Dr. Carl Werner, *Evolution: The Grand Experiment*, New Leaf Press, 2007, p. 125
[191] Barry Cox, "Mysteries of Early Dinosaur Evolution," *Nature*, 364:314, 1976
[192] Henry Gee, *In Search of Deep Time*, The Free Press, 2001, p. 113
[193] Alfred Sherwood Romer, *Vertebrate Paleontology*, 3rd edition, University of Chicago Press, 1966, p. 119
[194] Steve Parker, *Age of Dinosaurs*, Pegasus Publishing Ltd., 2000, p.116
[195] Alfred Sherwood Romer, *Vertebrate Paleontology*, 3rd edition, University of CHicago Press,1966, p. 126
[196] S.J. Gould, "A Darwinian Paradox", *Natural History*, January 1979, p. 40
[197] ___, *The Westbury Pliosaur*, Bristol City Museum and Art Gallery, 1989, p. 8
[198] Alfred Sherwood Romer, *Vertebrate Paleontology*, 3rd edition, University of Chicago Press, 1966, p. 126
[199] Kenneth R. Miller, Joseph Levine, *Biology*, Prentice Hall, 2002, p. 798
[200] *Ibid.*, p. 431
[201] Tom Kemp, "The Reptiles that Became Mammals," *New Scientist*, Vol. 92 March 1982, p.583

[202] Michael Denton, *Evolution: A Theory in Crisis*, London: Burnett Books, Ltd., 1985, p. 180
[203] Kenneth R. Miller and Joseph Levine, *Biology*, Prentice Hall, 2002, p. 434
[204] *Ibid.*, p. 821
[205] George B. Johnson & Peter H. Raven, *Biology*, Holt, Rinehart and Winston, 2004, p. 268
[206] Kenneth R. Miller & Joseph Levine, *Biology*, Pearson/Prentice Hall, 2008, p. 821
[207] *California Life Science*, Pearson Education, Inc., 2008, p. 294
[208] Roger Lewin, "Bones of Mammals' Ancestors Fleshed Out," *Science*, Vol. 212, No.4502, June 1981, p. 1,492
[209] Gordon R. Taylor, *The Great Evolution Mystery*, Harpercollins, 1983, p. 78
[210] M. J. Benton, *Vertebrate Paleontology*, 3rd ed. Blackwell Science, 2005, p. 359
[211] A.S. Romer, *Vertebrate Paleontology,* Third Edition, University of Chicago Press, 1966, p. 303
[212] *Ibid.*, pp. 304-310
[213] *Ibid.*, p. 308
[214] *Ibid.*
[215] *Ibid.*, p. 303
[216] William D. Schraer & Herbert J. Stoltze, *Biology: The Study of Life,* Prentice Hall, Fifth Edition, 1993, pp. 578, 580
[217] *Ibid.*, p. 580
[218] George B Johnson & Peter H. Raven, *Biology: Principles and Explorations,* Holt, Rinehart & Winston, 1996, p. 254
[219] George B. Johnson, *Biology Visualizing Life*, Holt, Rinehart & Winston, 1994, p. 181
[220] Eric Strauss & Marylin Lisowski, *Biology: The Web of Life*, 2nd edition, Scott Foresman-Addison Wesley, 2000, p. 238
[221] G.A. Kerkut, *Implications of Evolution*, Pergamon Press, 1960, p. 149
[222] H. Nilsson, *Synthetische Artbildung*, Verlag CWE Gleerup, 1954
[223] Steven M. Stanley, "Macroevolution and the Fossil Record," *Evolution*, vol. 36, May 1982, p. 464
[224] *Ibid.*
[225] Boyce Rensberger, *Houston Chronicle,* Nov. 1980, sec. 4, p. 15 -- reporting on a 4-day evolutionist conference held at Chicago's Field Museum
[226] J.B. Birdsell, *Human Evolution*, Rand McNally College Publishing Co., 1975, p. 169
[227] George B. Johnson & Peter H. Raven, *Biology: Principles and Explorations,* Holt, Rinehart & Winston, 1996, p. 254
[228] George G. Simpson, *The Major Features of Evolution,* Columbia University Press, 1953, p. 263
[229] J.T. Gregory, *University of California Publications in Geological Sciences*, vol. 26, 1942, p. 428
[230] Eric Strauss & Marylin Lisowski, *Biology: The Web of Life,* Scott Foresman-Addison Wesley, Second Edition, 2002, p. 238
[231] Bruce MacFadden, *Fossil Horses*, Cambridge University Press, 1994, p. 27
[232] David M. Raup, *Field Museum of Natural History Bulletin*, vol. 50 January 1979, p. 25
[233] George G. Simpson, *Life of the Past,* Yale University Press, 1953, p. 119
[234] G.A. Kerkut., *Implications of Evolution,* in International Series of Monographs in Pure and Applied Biology/Zoology Division, Vol. 4, Pergamum Press, 1965
[235] G.A. Kerkut, *Implications of Evolution*, 1960, p. 149
[236] George Gaylord Simpson, *Tempo and Mode in Evolution*, Columbia University Press, 1944, p. 105
[237] ___, *Science Explorer: Focus on Earth Science*, Prentice Hall, 2001, p. 395
[238] George B. Johnson & Peter H. Raven, *Biology,* Holt, Rinehart and Winston, 2004, p. 284
[239] *Ibid.*
[240] *Ibid.*
[241] Philip Gingerich, "Whales with Legs," *Science Digest*, November/December 1980, p. 25
[242] John H. Postlethwait & Janet L. Hopson, *Modern Biology*, Holt, Rinehart, & Winston, 2006, p. 302
[243] Philip D. Gingerich, N.A. Wells, D.E. Russell, & S.M. Shah, "Origin of Whales in Epicontinental Remnant Seas: New Evidence from the Early Eocene of Pakistan," *Science*, vol. 220, April 22, 1983, p. 405

[244] Neil A. Campbell, Brad Williamson, & Robin J. Heyden, *Biology: Exploring Life*, Pearson, 2009, p. 300
[245] E-mail interview with Lawrence G. Barnes, Curator of Vetebrate Paleontology at the Natural History Museum of Los Angeles County, California, conducted on December 21, 2005, by Dr. Carl Werner
[246] E. H. Colbert, *Colbert's Evolution of the Vertebrates*, Wiley-Liss, 2001, p. 392
[247] A.S. Romer, *Vertebrate Paleontology*, 3rd ed., University of Chicago Press, 1966, p 339
[248] E.J. Slijper, *Dolphins and Whales*, University of Michigan Press, 1962, p. 17
[249] J.G.M. Thewissen, S.T. Hussain, and M. Arif, "Fossil Evidence for the Origin of Aquatic Locomotion in Archeocete Whales," *Science*, 263(5144):210-212, January 14, 1994, Perspective by A. Berta, "What is a Whale?," same issue, p. 180-181
[250] A. Berta, "What is a whale?" *Science,* Volume 263, January 14, 1994, pp. 180-181
[251] Interview with Dr. Phil Gingerich, Paleontologist, University of Michigan, for video series *Evolution: The Grand Experiment*, conducted on August 28, 2001, by Dr. Carl Werner, p. 138
[252] Barbara J. Stahl, *Vertebrate History: Problems in Evolution*, McGraw-Hill, 1974, p. 489
[253] Dr. Carl Werner, Evolution: The Grand Experiment, New Leaf Publishing Group in conjunction with Audio Visual Consultants Inc., 2007, p. 145, quoting Dr. Philip Gingerich
[254] Interview with Dr. Phil Gingerich, Paleontologist, University of Michigan, for video series *Evolution: The Grand Experiment*, conducted on August 28, 2001, by Dr. Carl Werner
[255] Alfred Sherwood Romer, *Vertebrate Paleontology*, Third Edition, The University of Chicago Press, 1966, p. 309
[256] *Ibid.*
[257] *Ibid.*, 310
[258] *Ibid.*, p. 254.
[259] Holt, Rinehart and Winston, *Biology: Visualizing Life*, 1994, p. 176
[260] E.C. Olson, *The Evolution of Life,* Weidenfeld & Nicolson, 1965, p.180
[261] Philip J. Currie, *The Flying Dinosaurs*, Discovery Books, 1991, p. 32
[262] Kenneth Carpenter, "Monsters of the Sea and Air", found in *The Ultimate Dinosaur*, ed. by Byron Preiss and Robert Silverberg, Bantam Books, 1992, p. 240
[263] Carl Zimmer, "Masters of an Ancient Sky," *Discover*, February 1994, pg. 45
[264] Paul C. Sereno, "The Evolution of Dinosaurs," *SCIENCE*, vol. 284, June 25, 1999, p. 2143
[265] E.C. Olson, *The Evolution of Life*, Weidenfeld & Nicolson, 1965, p. 176
[266] Alfred Sherwood Romer, *Vertebrate Paleontology*, Third Edition, The University of Chicago Press, 1966, p. 338
[267] Carl Zimmer, "Into the Night," *Discover*, November, 1998, p. 114
[268] *Ibid.*, p. 112
[269] Edwin H. Colbert, Michael Morales, and Eli C. Minkoff, *Colbert's Evolution of the Vertebrates: A History of the Backboned Animals Through Time*, Wiley-Liss, 2001, p. 333
[270] George B. Johnson, *Biology Visualizing Life*, Holt, Rinehart & Winston, 1994, p. 210
[271] E.C. Olson, *The Evolution of Life,* Weidenfeld and Nicolson, 1965, p.180
[272] Pierre P. Grasse, *The Evolution of Living Organisms,* Academic Press, 1977, p. 30
[273] G.B. Johnson and P.H. Raven, *Biology: Principles & Exploration,* Holt, Rinehart, & Winston, 1996, p. 788
[274] Bernice Essenfeld, Carol Gontang, and Randy Moore, *Biology*, Addison-Wesley, 1994, p. 606
[275] G.B. Johnson and P.H. Raven, *Biology: Principles & Exploration*, Holt, Rinehart, & Winston, 1996, p. 738
[276] Raven & Johnson, *Biology*, 5th edition, McGraw-Hill, 1999, p. 314
[277] Alton Biggs, Chris Kapicka, Linda Lundgren, *Biology: The Dynamics of Life*, Glenco/McGraw-Hill, 2000, p. 386
[278] Kenneth R. Miller, Joseph Levine, *Biology*, Prentice Hall, 2002, p. 821
[279] *Ibid.*, p. 807
[280] George B. Johnson, Peter H. Raven, *Biology*, Holt, Rinehart & Winston, 2004, p. 268
[281] Kenneth R. Miller & Joseph Levine, *Biology*, Pearson/Prentice Hall, 2008, p. 807
[282] *Ibid.*

[283] *Ibid.*
[284] Elizabeth Coolidge-Stolz, et al., *Life Science*, Pearson Prentice Hall, 2007, p. 406
[285] W.E. Swinton in *Biology and Comparative Physiology of Birds,* ed. A.J. Marshall, Academic Press, 1960, Vol. 1, p.1
[286] W.K. Gregory, *New York Academy of Science Annals*, 1916, Vol. 27, p. 31
[287] S.J. Gould and Niles Eldredge, "Punctuated Equilibria: The Tempo and Mode of Evolution Reconsidered," *Paleobiology*, Vol. 3, 1977, p. 147
[288] Douglas Palmer, "Learning to Fly," *New Scientist,* vol. 153, March 1, 1997, p. 44
[289] Stewart Martin, *The Auk*, vol. 97, 1980, p. 86
[290] Paul Moody, *Introduction to Evolution*, 1970, pp. 196-197
[291] *ScienceNOW*, 8/4/2004
[292] E.C. Olsen, *The Evolution of Life*, Weidenfeld and Nicolson, 1965, p.82
[293] Peter Dodson, "International *Archaeopteryx* Conference," *Journal of Vertebrate Paleontology*, vol. 5, June 1985, p. 179
[294] Dr. Alan Feduccia, *American Scientist*, May/June, 1978 p. 302
[295] "What's a Hoatzin?" *Scientific American*, vol. 261, December 1989, p. 30
[296] Anonymous, "Bone Bonanza: Early Bird and Mastadon," *Science News*, vol 112, No. 13, September 24, 1977, p. 198
[297] J. Ostrom, *Science News, 112* (1977), p. 198
[298] Anonymous, "The Oldest Fossil Bird: A Rival for *Archaeopteryx*?" *Science*, Vol. 199, January 20, 1978, p. 284
[299] Alan Feduccia, "On Why Dinosaurs Lacked Feathers," *The Beginning of Birds*, Eichstatt, West Summary: Jura Museum, 1985 pp. 76, 199-284
[300] Alan Feduccia and Harrison B. Tordoff, "Feathers of *Archaeopteryx*: Asymmetric Vanes Indicate Aerodynamic Function", *Science*, Vol. 203, March 9, 1979, p. 1022
[301] Alan Feduccia, "Evidence from Claw Geometry Indicating Arboreal Habits of *Archaeopteryx*," *Science,* vol. 259, February 5, 1993, p. 792
[302] Larry D. Martin, "The Relationship of *Archaeopteryx* to other Birds," *The Beginnings of Birds*, M.K. Hecht, et al., 1985, p. 182
[303] Alan Feduccia, *The Origin and Evolution of Birds*, Yale University Press, 1996, p. viii
[304] *Ibid.*
[305] *Ibid.*
[306] *Ibid.*
[307] *Ibid.*
[308] *Ibid.,* p. ix
[309] George Johnson & Peter H. Raven, *Biology*, Holt, Rinehart & Winston, 2004, p. 268
[310] John Schwartz (quoting Feduccia), "China Fossil Discovery May Foul Birds-from-Dinosaurs Theory," *Las Vegas Journal and Las Vegas Sun,* November 5, 1995, p. 1C
[311] Douglas Futuyma, *Chronicle of Higher Education*, 10/25/96
[312] P. Shipman, *Taking Wing:* Archaeopteryx *and the Evolution of Bird Flight*, Simon and Schuster, 1988, p. 84
[313] A. Feduccia, et al., "Do Feathered Dinosaurs Exist? Testing the Hypothesis on Neontological and Paleontological Evidence," *Journal of Morphology*, 2005, 266:125-166
[314] Edward Stuart Russell, *The Diversity of Animals, an Evolutionary Study*, 2008, p. 118
[315] Kathy A. Svitil, "Plucking Apart the Dino-Birds," *Discover*, February 2003
[316] Charles Darwin, *The Origin of Species*
[317] George B. Johnson & Peter H. Raven, *Biology*, Holt, Rinehart & Winston, Inc. 2004, p. 283
[318] Niles Eldredge, *The Myths Of Human Evolution*, Columbia University Press, 1984, p.45-46
[319] Holt, 2004, Op. Cit.
[320] Kenneth R. Miller & Joseph Levine, *Biology*, Prentice Hall, 2002, p. 382
[321] Alton Biggs, Chris Kapicka, Linda Lundgren, *Biology:The Dynamics of Life*, Glencoe, McMillian/ McGraw-Hill, 1995, p. 431
[322] Albert Towle, *Modern Biology,* Holt, Rinehart, & Winston, 1993, p. 219
[323] Mark Ridley, "Who Doubts Evolution?" *New Scientist*, Vol. 90; June 25, 1981, p. 831

[324] Tom Kemp, "A Fresh Look at the Fossil Record," *New Scientist* Vol. 108; December 5, 1985, p. 67
[325] Stephen J. Gould, Lecture at Hobart & William Smith College, 4/2/1980
[326] Stephen Gould, "A Quahog is a Quahog," *Natural History,* V. 88, No. 7, 1979, pp. 18-26
[327] David Raup, "Conflicts between Darwin and Paleontology," *Field Museum of Natural History Bulletin*, vol. 50, January 1979, p. 25
[328] D.B. Kitts, *Evolution*, 28:467, 1974
[329] *Ibid.*
[330] Niles Eldredge, *Reinventing Darwin: The Great Debate at the Hight Table of Evolutionary Theory,* John Wiley & Sons, 1995, p.95
[331] D.S. Woodroff, *Science*, Vol. 208, 1980, p. 716
[332] Steven M. Stanley, "Resetting the Evolutionary Timetable," interview by Neil A Campbell, *Bioscience*, vol. 36, December 1986, p. 725
[333] David W. Swift, *Evolution Under the Microscope*, Leighton Academic Press, 2002, p. 295
[334] Steven M. Stanley, *Macroevolution: Pattern and Process*, W.M. Freeman and Co., 1979, p. 39
[335] D. Dwight Davis, "Comparative Anatomy and the Evolution of Vertebrates," in *Genetics, Paleontology and Evolution,* ed. Jepsen, Mayr and Simpson, Princeton University Press, 1974, p. 74
[336] G.G. Simpson, *The Major Features of Evolution*, 1953, p. 360
[337] James W. Valentine & Cathryn A. Campbell, "Genetic Regulation and the Fossil Record," *American Scientist*, November-December, 1975
[338] S.J. Gould, *Evolution's Erratic Pace,* Natural History, Vol. 86, May 1977, p. 14
[339] Stephen J. Gould, "The Return of Hopeful Monsters," *Natural History* 86:22, 1977
[340] Jeffrey H. Schwartz, *Sudden Origins,* John Wiley, 1999, p. 89
[341] Michael Denton, *Evolution: A Theory in Crisis*, Burnett Books, Ltd., 1985, p. 286
[342] Steven M. Stanley, "Macroevolution and the Fossil Record," *Evolution*, 36, No. 3, 1982, p. 460
[343] ______, "Is Man a Subtle Accident?", *Newsweek*, November 3, 1980
[344] Stephen J. Gould & Niles Eldredge, "Punctuated Equilibria: the Tempo and Mode of Evolution Reconsidered," *Paleobiology*, Vol. 3, 1977, pp. 145, 146
[345] *Ibid.*
[346] N.A. Takahata, "Genetic Perspective on the Origin and History of Humans," *Annual Review of Ecology and Systematics*, vol. 26, 1995, pg. 343
[347] Stephen M. Stanley, *The New Evolutionary Timetable,* Basic Books, 1981, p.77
[348] William D. Schraer and Herbert J. Stoltze, *Biology The Study of Life*, Prentice Hall, Fifth Edition, 1993, p. 604
[349] Colin Patterson, *Darwin's Enigma: The Fossil Record,* p. 100
[350] Stephen Jay Gould, "Opus 200," *Natural History,* 100, August 1991,: 16
[351] Steven M. Stanley, *The New Evolutionary Timetable*, Basic Books, 1981, p. 95
[352] Eldredge, quoted in a 1978 *Los Angeles Times*' article, "Alternate Theory of Evolution Considered," by George Alexander
[353] Stephen Jay Gould, "The Ediacaran Experiment," *Natural History*, Vol. 93, February, 1984, p. 23
[354] Stephen J. Gould, "Is a New and General Theory of Evolution Emerging?" *Paleobiology*, vol 6(1), January 1980, p. 127
[355] Niles Eldredge, quoted in "Alternate Theory of Evolution Considered," in Los Angeles Times, November 19, 1978.
[356] Stephen J. Gould, "Evolution's Erratic Pace," *Natural History*, vol. 86, May 1987, p. 14

HUMAN ORIGINS

Introduction

"Once upon a time..." That is how this section should start. Unfortunately for the world, the idea that humans evolved from apes or apelike ancestors is, for many, as sure as the facts of sunrise and sunset. Here is Earnst Mayr, one of the world's foremost speakers on the theory of evolution,

> "Since Darwin, every knowing person agrees that man is descended from the apes....Today, of course, there is no such thing as the theory of evolution; it is the fact of evolution..."[1]

In the past, amidst the various creationist/evolutionist arguments and debates over the topic of human origins, there has been an objection offered by the evolutionary speaker, "We don't say that humans evolved from apes. They evolved from apelike creatures." However, when you read the literature carefully, what you find from the top brass in evolution, such as Hooton of Harvard, is as follows,

> "If we are descended from apes our remote ancestors ought to look their part. You may not be willing to admit that you resemble an ape....But if that thousandth ancestor's forebearers become progressively more simian as you trace back the genealogical lines you will have to admit that somewhere in your family tree there squats an ape."[2]

Nevertheless, the high school textbooks prefer to say it in a way that is less brutal and more congenial, as shown below.

Textbook Evolution

> "...modern humans and apes evolved from a common apelike ancestor."[3] (Sylvia S. Mader, *Biology,* McGraw Hill)

> "Some *Australopithecus* species seem to have been human ancestors, while others formed separate branches off the main hominid line."[4] (Kenneth R. Miller & Joseph Levine, *Biology*, Pearson/Prentice Hall)

Tree Shrew

Textbook Evolution

> "According to the fossil evidence, the earliest primates were similar to the tree shrew shown above. Although the early primate's overall anatomy resembles that of a shrew, fossil evidence also shows some characteristics common to later primates."[5] (Eric Strauss & Marylin Lisowski, *Biology: The Web of Life*, second edition, Scott Foresman-Addison Wesley)

Science Versus

In his text on physical anthropology,, A. J. Kelso writes,

> "The transition from Insectivore to primate is not documented by fossils. The basis of knowledge about the transition is by inference from living forms."[6]

According to the theory of evolution, the first primates were primitive lemurs. However, the fossil record does not support this idea. Lemurs, according to Romer, appear "apparently as immigrants from some unknown area."[7]

Further, the old idea that tree shrews (tupaiids) are closely related to living primates is not supported by the fossil evidence either. Campbell, of the Indiana University, states,

> "I have attempted to indicate the large number of recent studies where results indicate that a close relationship between tupaiids and primates is unlikely."[8]

Anthropoids & Primates

Textbook Evolution

> "Early primates are thought to have evolved into two groups, the prosimians and the anthropoids. A prosimian...is a small, monkey-like primate with large, forward-facing eyes. Prosimian means "pre-monkey." Modern prosimians probably resemble the primates' earliest ancestors."[9] (Anthea Maton, *Evolution: Change Over Time*, Prentice Hall Science)

> "Most scientists agree that the earliest anthropoids evolved from a common prosimian ancestor some 50 million years ago."[10] (*Ibid.*)

> "Primates are fairly well represented in the fossil record, and the evolutionary history of primates has been reconstructed through fossils and comparative studies of modern primates."[11] (Alton Biggs, Chris Kapicka, Linda Lundgren, *Biology, The Dynamics of Life*, Glencoe, McMillian/ McGraw-Hill)

> "Figure 19.5 - All primates probably evolved from common ancestors. The evolutionary history of primates is based on the fossil record and comparative studies of living primates"[12] (*Ibid.*, p. 459)

> "Anthropoids are human-like primates that include monkeys, apes and humans."[13] (Anthea Maton, *Evolution: Change Over Time*, Prentice Hall Science)

Science Versus

With regard to the origin of primates in general, Elwyn Simons, a leading evolutionary paleoanthropologist, reports,

> "In spite of recent finds, the time and place of origin of order primates remains shrouded in mystery."[14]

Old World Monkeys

Textbook Evolution

"The other group of anthropoids evolved into Old World monkeys and hominoids....Old World monkeys live in Africa and Asia....Hominoids, which are cousins of the Old World monkeys, include apes---gorillas, gibbons, orangutans, chimpanzees---and humans.[15] (Anthea Maton, *Evolution: Change Over Time*, Prentice Hall Science)

"Old World monkeys evolved a little later in Africa, about 20 to 22 million years ago."[16] (Alton Biggs, Chris Kapicka, Linda Lundgren, *Biology, The Dynamics of Life*, Glencoe,McMillian/McGraw-Hill)

"Old World monkeys evolved more recently than New World monkeys."[17] (Alton Biggs, et. al., *Biology: The Dynamics of Life*, Glencoe Science, *National Geographic*, McGraw-Hill/Glencoe)

Science Versus

William Howles of Harvard states,

"As to the Old World Monkeys, even less is known of their past. But they too must go back to unknown Eocene ancestors...."[18]

Kelso also acknowledges that,

"Clearly the fossil documentation of the emergence of the Old World monkeys could provide key insights to the general evolutionary picture of the primates, but, in fact, this record simply does not exist."[19]

Robert Martin, the author of *Primate Origins and Evolution*, also emphasizes the lack of fossil evidence for the Old World monkey's ancestry,

"Overall, the fossil record can tell us very little about the early origins of Old World monkeys."[20]

New World Monkeys

Textbook Evolution

"A few million years after the split of prosimians and anthropoids, two anthropoid branches evolved. These branches, the two major groups of monkeys and apes, developed when the continents separated from one another. One group of anthropoids evolved into the monkeys found today in Central and South America. This group is called the New World monkeys."[21] (Anthea Maton, *Evolution: Change Over Time*, Prentice Hall Science)

"New World monkeys were the first modern anthropoids to evolve. According to the fossil record, Figure 19.5. New World monkeys evolved 30 to 35 million years ago."[22] (A.J. Kelso, *Physical Anthropology*, Lippincott Williams & Wilkins)

"The oldest monkey fossils are of New World monkeys are are 30 to 35 million years old. Although New World monkeys probably share a common anthropoidlike ancestor with the Old World monkeys, they evolved independently of the Old World monkeys because of geographic isolation. In Figure 16.6, you can see the worldwide geographic distribution of monkeys and apes.[23] (Alton Biggs, et. al., *Biology: The Dynamics of Life*, Glencoe Science, *National Geographic*, McGraw-Hill/Glencoe)

Science Versus

Kelso admits,

"The details of the Evolutionary background of the New World monkeys, the platyrrhini, would doubtless be informative and interesting but unfortunately we know very little about them."[24]

Martin refers to their origins as a mystery,

"The origins of the New World monkeys therefore remain a mystery, as the only substantial fossils are known from the late Oligocene and the Miocene, and by that stage they exhibit numerous characteristics clearly allying them with the modern forms. Indeed, without exception the substantial fossils that have been described resemble true New World monkeys (family Cebidae)"[25]

The famous evolutionary zoologist, William L. Straus Jr., recorded in his *Anthropology Today*,

"The phylogenetic history of the New World monkeys or platyrrhines is quite unknown."[26]

Apes

Textbook Evolution

"According to fossils found in Africa and elsewhere, apes evolved about the same time as Old World monkeys..."[27] (Alton Biggs, Chris Kapicka, Linda Lundgren, *Biology*, *The Dynamics of Life*, Glencoe, McMillian/McGraw-Hill)

"Fossils indicate that gibbons were the first apes to evolve, followed by orangutans and then African apes."[28] (*Ibid.*)

"Chimpanzees, gorillas, orangutans, and gibbons are modern examples of apes. Evidence indicates that the gibbons have been on earth the longest, around 35 million years."[29] (Eric Strauss & Marylin Lisowski, *Biology: The Web of Life*, second edition, Scott Foresman-Addison Wesley)

Science Versus

After reading the high school textbook, it all seems so cut and dried, with the gibbons first to arrive, followed by orangutans, etc. However, when you take an objective look at the evidence of fossils, the story becomes much less convincing.

Dr. Lyall Watson, anthropologist, comments,

> "Modern apes, for instance, seem to have sprung out of nowhere. They have no yesterday, no fossil record. And the true origin of modern humans--of upright, naked, tool-making, big-brained beings--is, if we are to be honest with ourselves, an equally mysterious matter."[30]

David Pilbeam, the Henry Ford II Professor of the Social Sciences at Harvard University, laments,

> "Unfortunately, the fossil record of pongids [apes] is non-existent, making a glaring deficiency in the whole story."[31]

Donald Johanson, the discoverer of *Lucy*, states,

> "At any rate, modern gorillas, orangs and chimpanzees spring out of nowhere, as it were. They are here today; they have no yesterday..."[32]

G.H.R. von Koenigswald, distinguished paleontologist and zoologist, comments regarding the gibbon,

> "(The gibbon) It's origin has not yet been traced."[33]

Henry Gee, senior editor of *Nature*, reports,

> "Fossil evidence of human evolutionary history is fragmentary and open to various interpretations. Fossil evidence of chimpanzee evolution is absent altogether."[34]

Australopithecines

In 1924 Dr. Raymond Dart, digging in Africa, discovered a small ape skull, which he decided to promote as one of man's earliest ancestors. He named this small skull *australopithecus*, meaning "southern ape." Dart proclaimed his find and his interpretation of its significance to the world in 1925 in the British journal, *Nature*. It soon became officially known as *australopithecus africanus*. Further, since it was found near the small township of Taung, it has also been referred to as the Taung child. The creature was estimated to have been about six years old. Unfortunately for Dr. Dart, the community of scientists oriented to the theory of evolution did not respond favorably to his ideas. This was in large part due to the fact that, at this time, Piltdown Man and the Peking Man fossils found in China had captivated the paleoanthropological community, so Dart's little ape skull was having to play a very distant second fiddle.

One of the main criticisms of Dart's fossil was that it was a juvenile. All paleontologists knew that the juvenile ape skull had a similar appearance to the human skull, but as the ape grew older the skull became increasingly apelike. With no adult specimen around to examine, Dart's ape remained unconvincing.

Later, John Robinson and Robert Broom found a cache of skulls in the Transvaal area of Africa. Some of the fossils were similar to that originally found by Dart. More importantly for Dart, one of the fossils was an adult version of his original find. Later, he found skeletons that were much more grossly apelike. In time, the ape fossils which were similar to Dart's fossil were named *Australopithecus Africanus*, while the ones more grossly apelike received the name *Australopithecus Robustus*. Since they were designated australopithecines, they were, at the beginning, thought to be significant ancestors which eventually led to "True Man."

Apart from Robert Broom, one of the first of the respected scholars in this field to show favorable attention to Raymond Dart's idea, was William King Gregory. In 1930 his article in *Science* magazine included the following statement,

> "If *Australopithecus* is not literally a missing link between an older dryopithecoid group and primitive man, what conceivable combination of ape and human characters would ever be admitted as such?"[35]

This is an interesting comment by Gregory seeing that at this present time, his boss, Henry Fairfield Osbourne, was championing Piltdown Man, which was eventually going to be shown to be a complete hoax. In 1938 Gregory came to Africa to visit Dart and to review the fossils. As a result of this, his confidence in the value of the australopithecines to the human lineage was strengthened. He was now fully on board with Dart. Gregory's position in the scientific community was high enough to help push *australopithecus* into a place for serious scientific consideration.

Robert Broom, it should be mentioned here, was indeed a gullible researcher, totally given to the theory of evolution. He was not only a supporter of the Piltdown man, but when he came to see Dart's australopithecine fossil for the first time, he walked across the room and fell to his knees in adoration of our apelike "ancestor."

Another famous detractor who spoke against the value of *australopithecus* to the human line was Sir Arthur Keith. In the 14 February 1925 edition of *Nature* he stated that Dart's fossil was most likely an ape. Actually, Dart's little ape was not really necessary to the evolutionary establishment at that time because of the attention being given to Piltdown Man, Nebraska Man and to Peking Man. However, things began to change for the worst for the theory of evolution with regard to human origins. Nebraska Man was shown to be the tooth of a pig, the Peking fossils had been totally destroyed by bombing during the first part of World War II, and eventually the famous Piltdown fossils began to wane in importance. The "evidence " for human origins from an apelike ancestor was dissolving decade by decade. In the light of this worrisome situation, Dart's apes became not only more interesting, but also more necessary.

So, though for a time, Dart's australopithecine fossils languished, receiving little professional attention, they were finally resurrected into acceptance and fame in large measure by LeGros Clark, a leading paleoanthropologist. In late 1946 he went to Africa and examined the Taung skull and other specimens. After reviewing the various fossils, he changed his mind with regard to the importance of the australopithecines to the human line and concluded that Dart was right after all and that they were the ancestors of true humans.

In January 1947 he presented his conclusions to the Pan African Conference on Prehistory which was being held in Nairobi, Kenya. There, Clark, one of the world's leading paleoanthropologists, referred to the australopithecines as hominids and not merely as apes. His remarks, supporting the importance of the australopithecines, appeared in the editorial pages of the 15 February 1947 issue of *Nature*, thus insuring them of worldwide coverage.

With this event, the dominoes began to fall. Immediately after the publication of the *Nature* article, Sir Arthur Keith wrote a letter to the journal's editor. In this letter he acknowledged that his assessment had been wrong and that Dart's had been correct. Thus, Dart's australopithecines were revived and given a new and very important status.

Louis Leaky, who was soon to become one of the world's leading field paleoanthropologists, never believed the australopithecines were ancestral to homo. Other important researchers also agreed with him. The high school textbooks' presentation of these "ancestors" and the evidence to the contrary will be coming up soon.

Ramapithecus

Along the way, a variety of candidates to fill the position of human ancestors have come and gone. One of the more dramatic specimens was that of *Ramapithecus*. The more fragmentary the fossilized remains, the easier it is for the evolutionist to make claims that the creature in question is truly an important hominid as opposed to an ape. To make the case for human evolution from apelike ancestors, it is imperative that the evolutionary establishment find and identify certain fossils as being "hominids." That is, creatures that were actually en route to becoming human beings. Nowhere is this idea better illustrated than in the case of *Ramapithecus*. The fossils of this specimen, which occupied the role of one of man's earliest apelike ancestors for almost 20 years, consist of a single upper jaw, which was broken in half in the middle, and a few teeth. Further, there was no palate. Since it was broken in the middle, this made it very difficult to discern whether it was originally U shaped, as in apes, or parabolic shaped, as in humans.

Actually, the original fossils had been discovered in 1932. In that year, G. Edward Lewis, a doctrinal student at Yale, discovered these fossils during an expedition to the Siwalik Hills in India. He was the one who named these fossils *Ramapithecus*, or "Rama's Ape." In 1934 he published an article in the *American Journal of Science.* In that article he declared this creature to be a very early hominid. However, Alex Hrdlicka of the Smithsonian, who was at that time one of the most prestigious scientists in the area of anthropology, did not agree with Lewis's assessment. He also wrote an article in the *American Journal of Science*, which was very critical of Lewis's claims. As a result, Lewis's claims were not taken seriously by the paleoanthropological community, and the fossils were soon forgotten.

However, in 1961 Elwyn Simons came to teach at Yale and was able to review the original *Ramapithecus* fossils and soon came to the conclusion that it had certain features that qualified it as an early human ancestor. In his opinion the individual cheek teeth were humanlike, and most importantly, the design of the jaw was similar to humans and not to apes. In his 1961 paper he stated that the jaw, "can be determined as parabolic [like an arch] and not U-shaped."[36] To him, the two jaw pieces were to be arranged to have a parabolic configuration rather than a U-shaped design as is common for apes.

Later, in Simons' paper to *Anthropology* he wrote,

> "*Ramapithecus* Punjabicus is almost certainly man's forerunner of 15 million years ago. This determination increases tenfold the approximate time period during which human origins can now be traced with some confidence."[37]

For the next ten years he and his colleague, David Pilbeam, who came to Yale in 1963, co-authored papers which maintained Rama as a true hominid. They were of the opinion that *Ramapithecus* walked on two legs and even made and used tools. This time the news about Ramapithecus was welcomed into the paleoanthropological community. Even Raymond Dart was pleased with their work. So, as a result of Simon and Pilbeam's work, *Ramapithecus*, as a hominid, rose to prominence and seemed to firmly support the evolutionary contention that human origins began with very ancient, apelike creatures. This view held sway throughout the 1960's and much of the 1970's.

Along the way, one of the events which began *Ramapithecus's* downward fall occurred in 1961. A problem that had arisen in paleoanthropology was that there were too many species for hominids. Everyone that found anything wanted to name his own species. The names were beginning to pile up due to this careless means of identification. Simons and Pilbeam decided to reorganize these fossils by lumping ones that were quite similar into one species. This work, which was welcomed by many, significantly reduced the number of separate species. However, this did not go over well with Louis Leakey because a couple of his important fossils got lumped into a group which essentially caused them to lose their importance. One of them was a fossil of a supposedly 15 million-year-old creature, which Leakey had named Kenyapithecus Wickeri. Louis Leakey had considered his fossil, which consisted of an upper jaw, to be an early hominid like *Ramapithecus*. In fact, he found his fossil in the same year (196l) that the once abandoned *Ramapithecus* had been raised to the new status of hominid by Simons. Leakey had been sure that his fossil was different from that of *Ramapithecus* and that it was older (and thus more important). However, Simons and Pilbeam studied the two jaws and concluded that they were essentially the same. Thus, since *Ramapithecus* had been discovered first, Leakey's name Kenyapithicus Wickeri would have to be rejected. Later, at a meeting consisting of a variety of men involved in this field of study, including Louis Leakey, Pilbeam made a presentation, in which he made the statement that Kenyapithecus Wickeri (Leakey's fossils) was Ramapithecus Punjabus. While this nomenclature stuck it was not, in the long run, going to be helpful to the hominid status of *Ramapithecus*.

The chips began to fall against this claim in the early 1970's when the work of a British student initiated a rejection of the accepted view that the jaw shape of *Ramapithecus* was parabolic as in humans. At this time, Peter Andrews was working with Louis Leakey in Kenya. Andrews, who had earned a degree in anthropology from Cambridge University, was allowed in 1970 to examine the original fossils of Kenyapithicus Wickeri. Further, he was also able to review a lower jaw which had been found in the same area. The lower jaw had already been identified by both Leakey and Simons as having belonged to an ape. When Andrews compared the two jaws (upper and lower) he found that they fit together perfectly.

As a result of this discovery, Andrews wrote an article for *Nature* in May of 1971 in which he stated that Kenyapithicus now had a lower jaw. This statement had serious implications for Simons' and Pillbeam's *Ramapithecus* status. This is because Kenyapithicus Wickeri had been touted as a putative hominid, that is, as an ape that had incipient human features. Since the lower jaw was recognized to be that of an ape, this should have led to the conclusion that *Ramapithecus* was just an ape. However, it was necessary to keep Rama alive as a hominid, so they simply said that Rama was a much more "primitive" hominid than they had expected.

Afterwards, Andrews and Walker, a paleoanthropologist and good sculpture, got together and, working from plaster replicas, made a new reconstruction of the Kenyapithecus jaws. Their reconstructed jaw shape was much different from that which Simons made for Ramapithecus. In their reconstruction, the jaws did not have the curved tooth rows as are found in humans. It is important to remember that, according to Simons and Pilbeam, Kenyapithecus is simply another fossil of *Ramapithecus*.

In 1980 in Turkey, and in 1982 in Pakistan, parts of *Sivapithecus* faces were discovered. In reviewing the fossils, the paleontologists were able to see that this extinct ape was closely related to the orangutan. It has also been acknowledged that *Ramapithecus* was closely related to *Sivapithecus*. Hence, Rama is closely related to the orangutan and is therefore not a hominid.

In January 1982, Pilbeam wrote a paper regarding the new *Sivapithecus* face. It described the face well, but was quite tepid in relating Siva to Rama and what the consequences might be for Rama's hominid status. Afterwards, Andrews was invited by *Nature* to write another paper commenting on Pilbeams's paper. This paper included a straightforward assessment of the significance of the new *Sivapithecus* fossils,

> "It thus appears that *Sivapithecus* (including '*Ramapithecus*') is part of the orang-utan clade....In other words '*Ramapithecus*' can no longer be considered as part of the human lineage."[38]

In other words, for about 20 years the kids in the high schools and colleges were informed that *Ramapithecus* was solid evidence that humans evolved from apelike creatures. Most of them have probably never learned that *Ramapithecus* was merely an orangutan.

Elwyn Simons states categorically,

> "*Ramapithecus* is disqualified as a hominid because it is like *Sivapithicus*, and *Sivapithicus* has oranglike features.[39]

After the *Ramapithecus* fiasco was more or less concluded, Pilbeam made some interesting comments which acknowledge the subjectivism in this quest for human origins from the evolutionary point of view. These are remarkable admissions by Dr. Pilbeam, which actually could be applied to the entire set of so-called hominids, such as the australopithecines,

> "I will never again cling so firmly to one particular evolutionary scheme. I have come to believe that many statements we make about the hows and whys of human evolution say as much about us, the paleoanthropologists and the larger society in which we live, as about anything that 'really' happened."[40]

> "...perhaps generations of students of human evolution, including myself, have been flailing about in the dark; that our data base is too sparse, too slippery, for it to be able to mold our theories. Rather the theories are more statements about us and ideology than about the past."[41]

With regard to *Ramapithecus*, Roger Lewin, in his very informative book, *Bones of Contention*, again acknowledges the subjectivism involved in the construction of such fragmentary evidence into a hominid of commanding importance.

> "In this chapter is recounted the enthusiasm with which meager fossil evidence was interpreted--or more properly, overinterpreted--as a sure sign of incipient humanity..."[42]

A. Africanus

Introduction

With regard to the evolution of humans and the concept of hominids, the Holt, Rinehart, & Winston text states,

> "Paleontologists continue to find new hominid fossils. Although the exact classification of some fossils is strongly debated, it is clear that human evolution did not proceed as a single lineage of increasingly humanlike forms. Rather, several hominid forms arose, thrived, and

became extinct over the past 7 million years, as shown in Figure 43-20. Furthermore, different species of hominids may have coexisted in time and possibly interacted."[43]

The paragraph above represents a rather bland, though strongly evolutionary, presentation of the hominid evolution. For many years there was much dogmatism as to which fossil hominids came first, second, and third. However, the discovery of more fossils, as the reader will see toward the end of this chapter, has thrown a great deal of confusion into the idea of human evolution. Consequently, the more recent biology texts have been inclined to present less dogmatic statements as to the order of the line of hominids eventuating in human evolution.

A. Africanus has, for a long time, been the star performer in the imaginary history of humanity. The first fossil, as mentioned earlier, was found by Dr. Raymond Dart in Africa. Below are the current beliefs regarding *A. Africanus* as put forth in the state adopted high school biology textbooks.

Textbook Evolution

"Figure 19.7 The Taung child was the first australopithecine ever discovered. The braincase, face, and teeth are chimplike, but the position of the foramen magnum indicates that it walked upright like humans."[44] (Alton Biggs, et. al, *Biology,* Glencoe)

With regard to the evolutionary status of *A. africanus*, the Holt text reports,

"Dart argued that *A. africanus* was the direct ancestor of humans, the long sought "missing link" between apes and humans. Most scientists at the time dismissed Dart's discovery as the skull of a young ape. Yet many *A. africanus* fossils have been discovered since 1924, and we now know that this species was a hominid, not an ape."[45] (George B. Johnson, *Biology: Visualizing Life*, Holt, Rinehart & Winston)

"*Australopithecus africanus* was bipedal and existed from 2.5 million to 3 million years ago. Whether *A. africanus* was a direct ancestor of humans is still under debate."[46] (*Ibid.*)

"A possible descendant of Lucy's species was *A. africanus*, which lived about 2.3 to 3 million years ago."[47] (John H. Postlethwait & Janet L. Hopson, *Modern Biology*, Holt, Rinehart, & Winston)

Science Versus

Charles Oxnard, who was previously the Dean of Graduate School and professor of Biology and Anatomy at USC, conducted a multivariate analysis comparing the pelvic bones of the australopithecine, apes, and man. His conclusion was that the australopithecine had a mode of locomotion similar to an orangutan. The first three quotes below are Oxnard's statements,

"In one sense you may think there is no problem. For most anthropologists are agreed that the gracile australopithecines... are on the main human lineage....This is the view that is presented in almost all textbooks....However, anatomical features in some of these fossils provide a warning against a too-ready acceptance of this story..."[48]

"Yet another view might suggest that the fossil arose from the African apes via modern humans!"[49]

"We may well have to accept that it is rather unlikely that any of the australopithecines, including "Homo habilis," can have had any direct phylogenetic link with the genus Homo."[50]

J.B. Birdsell, author of *Human Evolution*, who taught anthropology at UCLA until his retirement, concludes,

"It is now evident that neither *Australopithecus africanus* nor robustus was an ancestor of modern man."[51]

It is important that the high school students not allow themselves to be easily convinced, even though some authors are quite dogmatic in their assertions, such as physical anthropologist, Kenneth Oakley, is with regard to the tool-making capacity of the Australopithecines,

"There no longer seems any reasonable doubt that the Africanus Australopithecines were hominids, that some of them made tools according to a definite tradition and therefore, although small-brained, were 'men' in the sense of regular tool making."[52]

This tool-making capacity is, in fact, subject to serious dispute, as award winning science book writer, David Lambert, declares,

"Experts dispute its ability to manufacture tools: most bone 'tools' found near africanus fossils have proved to be remains of meals consumed by hyenas or other carnivores."[53]

A. Robustus

Introduction

Probably the most famous of the robustus fossils was that found by Louis Leakey's wife, Mary, in 1959 in Kenya, Africa. One of Louis Leakey's financial benefactors was Charles Boice. Consequently, Louis named this robustus specimen *Zinjanthropus boisei*. When Louis first saw the fossil he was a bit disappointed because at first glance it appeared to be just another australopithecine. However, Leakey dearly wanted to find a fossil which would, in his opinion, be very important to the evolutionary development of humans. So, he began scrutinizing the skull which had just been found. The more he observed the skull, the more it appeared to him to have unique aspects which removed it from the australopithecine group or "near men" as he would call them.

Textbook Evolution

"Subtitle: Other australopithecines have been discovered."
"A stockier kind of australopithecine was unearthed in South Africa in 1938. Called *A. robustus*, this species had massive teeth and jaws. In 1959 the fossil hunter Mary Leakey discovered the even more solidly built *A. boisei*, shown in Figure 11.11, in Tanzania. Nicknamed 'nutcracker man' because of its massive jaws. *A. boisei*

had a large, bony ridge running along the crest of the head to anchor powerful jaw muscles. Excavations in 1895 turned up yet another australopithecine, a massively boned species resembling *A. boisei.*"[54] (George B. Johnson, Holt, Rinehart, & Winston, *Holt Biology: Visualizing Life*)

"Figure 14-8 because of an incomplete fossil record, scientists differ in their interpretations of how australopithecines evolved. Some researchers would put *A. robustus* on one branch, while others would put the same species on another branch."[55] (George B. Johnson and Peter H. Raven, *Biology: Principles and Explorations*, Holt, Rinehart and Winston)

Science Versus

As mentioned earlier, the most famous of the robustus fossils was that found by Mary Leakey in 1959. After taking a good look at the new fossil which Mary had found, Louis was a bit disappointed and related to her that it was just another australopithecine. He wanted to find something that was more significantly related to the ancestry of true man. However, as he began to review the skull from all angles, it was not long before he convinced himself that this was not just another australopithecine, but something that was very important to human evolution. Soon, he had written up a paper which he sent to Nature, making the case that this creature was definitely on the direct line to modern man.

Really and truly, it was just another of the robustus type specimens, but Leakey had persuaded himself otherwise. Not everyone else, however, was eager to take to share his opinion. A good friend of his, F. Clark Howell, was the first scientist to see Zinj after Mary and Louis had dug it up and examined it. What did he think of it? To him it looked like another robust australopithecine. He was not alone, as Lewin reveals in *Bones of Contention*,

"Most of the paleoanthropological establishment were, however, more than a little skeptical of Leakey's various claims, including the idea that Zinj was anything other than an australopithecine, and that he was on the direct line to man."[56]

Nevertheless, Louis Leakey brought his fossil to the 1959 Darwin Centennial where it was a big hit. From there on, Leakey began to get much financial support for his anthropological work in Africa. Lewin reports,

> "In retrospect, it is easy to see that Leakey's eagerness to find an early ancestor to 'true man' had led him to overinterpret the anatomical evidence."[57]

Louis Leakey argues that his newly found skull is significant,

> "At first sight, this new skull very strongly recalls the 'ape-men'...that Broom and Robinson found at Swartkranz, in the Transvaal....But when the skull is closely compared with the Swartkranz specimen known as *[Australopithecus robustus]*, it is found that the differences are far more numerous--more significant than the resemblances."[58]

Supposedly, there were primitive stone tools found in association with the Zinj fossils. Leakey wanted to consider his find more advanced than the average Australopithecine, and thus insisted that his creature was the hunter and not the hunted,

> "There is no reason whatever, in this case, to believe that the skull represents the victim of a cannibalistic feast..."[59]

How did Zing get knocked out of the limelight? It was done by Leakey himself, as Lewin reports,

> "*Zinjanthropus*, by the way, was relegated to *Australopithecus* by Leakey in the *Nature* paper that announced *Homo habilis*. Its anatomy had not altered, but Leakey's claim for its human status had."[60]

And what about all of this business that the robust australopithecines were important to the process of human evolution? Actually, the evidence shows that they were not the hunters, they were the hunted. The ancient apes, now referred to as *A. robustus*, appear to have been the meals of other creatures. Lambert (and the Diagram Group) informs us,

> "All undisputed robustus fossil finds come from South African caves, where the creatures' carcasses were evidently dragged or dropped by carnivores."[61]
>
> "Before then it probably gave rise to boisei, although many experts identify that with this species. Skull comparisons have even led some writers to suggest that robustus was ancestral to the gorilla."[62]
>
> "Boisei resembled a more massive version of Australopithecus robustus, from which it probably evolved, although many paleoanthropologists consider it just a regional variant of that species."[63]

The ancient apes, currently referred to as *A. robustus*, have been shelved as credible ancestors of humans. Boyce Rensberger, writing in *Science* 84 relates,

> "The other, *A. Robustus* (a closely related form is sometimes called *A. Boisei*), thrived for almost a million years only to come to an evolutionary dead end."[64]

Referring to *A. Robustus* and *A. Boisei*, the high school textbook, *Modern Biology*, also agrees that neither of these were ancestral to humans,

> "The general appearance of these other australopithecines suggests that while they may have descended from *A afarensis* (Lucy's species), they were probably not ancestral to modern humans."[65]

Homo habilis

Introduction

While Louis Leakey was still huffing and puffing about the importance of Zinjanthropus to the human lineage, his son Jonathan in November 1960, began finding some new fossils in the Olduvai Gorge in Kenya at a location very near where Zinj was found. These fossils represented the remains of a creature judged to have been about 12 years of age. Further, according to Leakey's reckoning, the specimen lived in an era slightly earlier than did Zinj. So, from the evolutionary standpoint, it preceded Zinj in time. Consequently, it soon became known as the pre-Zinj child.

Again, Louis Leakey came very quickly to the conclusion that this was indeed not merely an ape, but a hominid. In fact, it was a very important hominid and deserved to be considered a new species. Now, after working so hard to convince the big guns in the field of paleoanthropology of the importance of Zinj, he had to start all over again making an even bolder assertion for the new fossil.

On 4 July 1964 the big announcement came in the pages of *Nature* that Louis Leakey was now describing a brand new species, which was extremely important to our understanding of the evolutionary development of homo sapien. The immediate result was voluminous opposition.

At first, his friend, Phillip Tobias, only conceded in a letter to the British magazine, *Discovery,* that the fossils of *Homo habilis* "represent a stage in the development of man which neatly closes the gap between the most advanced *Australopithecus* and the lowliest *Homo.*"[66]

Leakey dismissed this idea and firmly stated in his final letter to *Discovery*, "I believe that *Homo habilis* represents a distinct branch of *Homo* possibly leading to *Homo sapiens,* and do not see it as an intermediate link between *Australopithecus* and *Homo erectus.*"[67]

Louis Leakey had no confidence that *homo erectus* types or *australopithecines* were the direct ancestors of homo sapien.

Eventually, Phillip Tobias went along with Leakey's view of the importance of homo habilis, and in 1964 the two of them concluded in an issue of *Nature* that the *Homo habilis* fossils should be assigned to the genus *Homo*, that is, 'true man,'

> "We have come to the conclusion that, apart from *Australopithecus* (*Zinjanthropus*), the specimens we are dealing with from Bed I and the lower part of Bed II at Olduvai represent a single species of the genus *Homo* and not an australopithecine,"[68]

Textbook Evolution

> "(Figures 11.15 b) *Homo habilis* 1960s, Olduvai Gorge, Tanzania. Louis Leakey discovered stone tools, crushed bones,and parts of a skull with a brain volume much larger than any australopithecine brain. These remains belonged to the species *H. habilis*. Importance: These were the earliest *Homo* fossils yet discovered, although many scientists doubted Leakey's identification of the skull"[69] (George B. Johnson, *Biology: Visualizing Life*, Holt, Rinehart, & Winston)

> "Because of its association with tools, this early human was named *Homo habilis*. The name *Homo habilis* is from the Latin *homo*, meaning "man," and *habilis*, meaning "handy"....*H. habilis* lived in Africa for 500,000 years and then became extinct, replaced by a new species of human with an even larger brain."[70] (George B. Johnson & Peter H. Raven, *Biology: Principles and Explorations*, Holt, Rinehart, and Winston)

> "Scientists generally agree that *Homo habilis* is a direct ancestor of our modern human species, *Homo sapiens*."[71] (Eric Strauss & Marylin Lisowski, *Biology: The Web of Life*, second edition, Scott Foresman-Addison Wesley)

> "Unlike australopithecine fossils, these newly discovered remains were found along with stone tools. This finding led to the naming of the new species as *Homo habilis*....Tool marks on animal bones found near the hominid fossils suggest that *H. habilis* ate meat."[72] (*Modern Biology*, Holt, Rinehart, & Winston)

> (Regarding *Homo Habilis*) "Importantly, the new fossils were found along with stone tools. Scientists named the new species *Homo habilis*, the "handy human."[73] (Alton Biggs, et. al., *Biology: The Dynamics of Life*, Glencoe Science, *National Geographic*, McGraw-Hill/Glencoe)

Science Versus

Louis Leakey found *Homo habilis* in the Olduvai Gorge in Tanzania. After a careful review of the skeleton, he decided that this creature was to be put in the same genus as man. This was the earliest species to be given the name homo. It was thought to be about the same size as *Homo erectus* and it was claimed that *Homo habilis* made tools. However many paleontologists disagreed with this special categorization for *Homo habilis*. They contended that it was just another *australopithecine*.

Homo habilis was supposed to have been intermediate between *A. afarensis* and *Homo erectus* and, therefore, it was assumed that when a more complete skeleton was found, its size would show it to be truly an intermediate species. However, when the more complete fossil of *Homo habilis* was found, it was hardly any larger than the tiny fossil of *A. afarensis*, Lucy. Henry McHenry, professor of anthropology at the University of California at Davis, comments,

> "Because *Homo habilis* is considered to be an evolutionary intermediate between the relatively small *Australopithecus afarensis* and the relatively large *Homo erectus*, everyone assumed *Habilis* would be of intermediate size."[74]

He goes on to say that the discovery of the more complete skeleton of *Homo habilis* was a distinct disappointment,

> "The discovery at Olduvai of arm and leg bones of the first member of the genus *Homo* shows that the creature was much more primitive than had been expected."[75]

Though in size it was definitely disappointing as to what an "advanced" *australopithecine* should be, Donald Johanson, of the Institute of Human

Origins in Berkeley, California affirms, "There's no doubt that it is *Homo habilis*."[76]

The newly discovered fossils of *Homo habilis* showed that it was definitely not in size half-way between *A. afarensis* and *Homo erectus*,

> "I laughed when I saw it," says Henry McHenry. "It is so tiny"[77]
>
> *Homo habilis*, actually, was shown to be the same size as an ordinary monkey.
>
> "McHenry states: 'Now we have to go back to the fossil collections and fish out the hominid limb bones that have been misclassified as monkeys.'"[78]

In other words, if *Homo habilis*, as small as it is, is to be considered a hominid, than they are going to have to review their many boxes of monkey fossils. Who knows how many "hominids" they may find there?

Both Ian Tattersall of Columbia University and C. Loring Brace of the University of Michigan have gone on record as stating that *Homo habilis* has lost it's definitive character,

> "It is increasingly clear that *Homo habilis* has become a wastebasket taxon, little more than a convenient recipient for a motley assortment of hominid fossils from the latest Pliocene and earliest Pleistocene..."[79]
>
> "*Homo Habilis* is an empty taxon inadequately proposed and should be formally sunk."[80]

A. Afarensis (Lucy)

In 1976, Donald Johanson, discovered two separate sets of fossilized bones. In sight 333 the fossils of human hand, foot, and leg bones were discovered. In site 162 the remains of a 70 pound ape skeleton was unearthed. Johansen reported them in 1976 as being two separate species. However, due to the influence of his assistant, Tim White, in 1979 he announced that they actually represented one species. This was a remarkable announcement inasmuch as the one set was totally human, the other set was totally ape, and the two sets were actually found 1,000 miles apart.

Textbook Evolution

> "Since Johanson's discovery, many other specimens of A. *afarensis* have been unearthed, some of which are estimated to be 3.9 million years old. Most researchers agree that these smaller A. *afarensis* individuals represent the true base of the human family tree. They appear to be the first members of the genus *Australopithecus.* As shown in Figure 14-8, they are probably ancestors of all the other australopithecines as well as of the genus *Homo*, which includes our species."[81] (George Johnson and Peter Raven, *Biology Principles and Explorations,* Holt, Rinehart and Winston)
>
> (Regarding Lucy) "The fossil primate's height ranged from 1 to 1.5 m, probably varying according to gender. The pelvis and leg bones, however, clearly indicated that the fossil organism was an upright-walking hominid."[82] (*Modern Biology*, Holt, Rinehart, & Winston)
>
> "The best known species is *Australopithecus afarensis*--described from a remarkably complete female skeleton, nicknamed Lucy, who stood only about 1 meter tall."[83] (Kenneth R. Miller & Joseph Levine, *Biology*, Pearson/Prentice Hall)
>
> "One early group of hominids, members of the genus *Australopithecus*, lived from about 4 million to a million years ago....Some *Australopithecus* species seem to have been human ancestors, while others formed separate branches off the main hominid line."[84] (*Ibid.*)

Science Versus

A common sense interpretation of the facts easily leads to the conclusion that these two separate fossil finds represent two species and not one. Author, William Fix, relates in his book, *The Bone Peddlers*, that,

"Mary Leakey has opted for two species complaining that the fossils come from separate sites a thousand miles apart..."[85]

According to *Science News*,

"The discovery of WR 17,000, the black skull, supports the idea that *A. afarensis* is really two species. Other researchers, however, say that WT 17,000 supports their previous contentions that the hadar fossils represent two different species- a slender, gracile type, exemplified by Lucy...and a robust type."[86]

Richard Leakey comments on the extreme fragmentary nature of Lucy's skull. From an article in *The Weekend Australian*, we read,

"Echoing the criticism made of his father's *habilis* skulls, he added that Lucy's skull was so incomplete that most of it was 'imagination made of plaster of Paris', thus making it impossible to draw any firm conclusions about what species she belonged to."[87]

There are many paleontologists who do not consider Lucy as an ancestor of man. They, like Adrienne Zihlman of the University of California at Santa Cruz, acknowledge that,

"Lucy's fossil remains match remarkably well with the bones of a pygmy chimp."[88]

Lucy (A. afarensis) is considered to be the earliest hominid in the line leading to man, dated at about 3.8 million years old. The new Homo habilis (OH 62), an adult female, is 3.5 feet tall. It is essentially the same size as Lucy. Furthermore, the post-cranial skeleton is as primitive and ape-like as that of Lucy. The fossilized arms of this new OH are long and ape-like. Like all the other australopithecines it shows itself to be just another tree swinger, having long arms and curved fingers. However, OH 62 is dated to be 1.8 million years old. This means, of course, that between Lucy and OH 62, there was 2 million years of no evolutionary change.

Actually, Lucy was an ape and not a transitional species between apes and humans.

First, as Lambert reports, Lucy looked like an ape,

"The creature looked like a small yet upright chimpanzee."[89]

Second, Lucy's face was that of an ape,

"The face was apelike with a forehead, browridge, flat nose, no chin but jutting jaws with massive back teeth."[90]

Third, as reported in *Nature*, Lucy had the toes of an ape,

"A. afarensis....The recent description of four articulating foot bones from 3-3.5 million years deposits in the South African cave site of Sterkfontein support this....The divergent big toe indicates some degree of prehensile grasping as in apes. Developmental patterns were also more ape-like than human...ecologically they may still be considered apes."[91]

Fourth, regarding the pelvis, toes, and finger bones, William Howells, professor of anthropology at Harvard observes,

"...the pelvis was by no means modern nor were the feet: the toes were more curved than ours; the heel bones lacked our stabilizing tubercles; and a couple of small ligaments that, in us, tighten the arch from underneath, were apparently not present. The finger bones were curved as they are in tree-climbing apes."[92]

Fifth, Lucy looked like a chimpanzee. Adrienne Zihlman states,

"Zihlman compares the pygmy chimpanzee to 'Lucy,' one of the oldest hominid fossils known and finds the similarities striking. They are almost identical in body size, in stature; and in brain size.... These commonalities, Zihlman argues indicate that pygmy chimps use their limbs in much the same way Lucy did...."[93]

The high school textbook, *Biology: Visualizing Life*, tells the students,

"Since the discovery of Lucy, many other fossils of *A. afarensis*, some over 3.5 million years old, have been unearthed in Ethiopia and in Tanzania. Since some of these fossils were found near the site of the bipedal footprints shown on the previous page, it is possible that *A. afarensis* individuals left these footprints."[94]

Were the australopithecines indeed capable of bipedal locomotion? When did it actually begin in the hominid line? Here we have more presumption than fact for, as Brian Richmond (Department of Anthropology, George Washington University) and David Strait (Department of Anthropology, University at Albany) report in *Nature*, the origin of bipedalism remains an unsolved problem,

> "Fossil evidence demonstrates that by 4.1 million years ago and perhaps earlier, hominids exhibited adaptations to bipedal walking. At present, however, the fossil record offers little information about the origin of bipedalism, and despite nearly a century of research on existing fossils and comparative anatomy, there is till no consensus concerning the mode of locomotion that preceded bipedalism."[95]

D. Schmitt of Duke University also comments,

> "...because of the paucity of the fossil record, the fragmentary nature of fossil remains, and the difficulty of inferring behavior from fossils, significant questions remain unanswered concerning the evolution of human bipedalism."[96]

Footprints of Humans or Ape-man (Lucy)?

Textbook Evolution

> "In 1978, Mary Leakey made the dramatic discovery of several sets of bipedal footprints in volcanic ash at Laetoli, a site located south of Olduvai Gorge. When the ash was analyzed, the footprints were dated at 3.6 million years old, clearly demonstrating that human ancestors became bipedal very early."[97] (*Modern Biology*, Holt, Rinehart, & Winston)

Though the footprints are clearly human-like footprints, the textbook tells the readers that they were made by human ancestors which "became bipedal very early." In other words, they were ape-men of some sort. The Holt text "clarifies" that these footprint makers represented an ape-man group that diverged "from the chimpanzee line,"

> "The footprints, shown in Figure 11.10, were preserved when the ash hardened. They reveal small but very humanlike feet, lacking the ape's opposable toe. Our ancestors or very close relatives walked upright only 1.5 million years after diverging from the chimpanzee line."[98] (George B. Johnson, *Biology: Visualizing Life*, Holt, Rinehart, & Winston)

Miller and Levine state below what others have also stated with regard to the human footprints found by Mary Leakey. It has been repeated over and over that they were made by people, not unlike ourselves. Yet, because of the presumption of evolutionary ancestors, there is an unwillingness to assume the obvious--that they were made by humans. Instead, in their thinking, they must have been made by "hominids" with human-like feet,

> "The humanlike footprints shown in Figure 32-17, which are between 3.8 and 3.6 million years old, were probably made by members of the same species as Lucy."[99] (Kenneth R. Miller & Joseph Levine, *Biology*, Pearson/Prentice Hall)

Science Versus

The dating of these tracks (3 million years plus) suggests to the evolutionist that they must have been made by some kind of an australopithecine. However, direct observation shows the footprints to be entirely human even though some have, in their artwork, given the creatures a very ape-man face. *TIME* magazine reported,

> "...They belonged to the genus Homo (or true man) rather than to man-apes (like Australopithecus, who was once thought to be the forerunner of man but is now regarded as a possible evolutionary dead end)...they were 3.35 million to 3.75 million years old....They would, in Mary Leakey's words, be people 'not unlike ourselves,'..."[100]

Russell Tuttle of the Department of Anthropology at the University of Chicago, further identifies these tracks,

> "...the 3.5 million-year-old footprint trails at Laetoli site G resemble those of habitually unshod modern humans. None of their features suggest that the Laetoli homnids were less capable bipeds than we are."[101]

Let us review again, as stated above, the party line--that these human-like footprints were made by a creature similar to Lucy (*Australopithecine*),

> "The humanlike footprints shown in Figure 32-17, which are between 3.8 and 3.6 million years old, were probably made by members of the same species as Lucy."

However, researchers, such as Tuttle, dismiss the idea that these footprints were made by creatures such as Lucy,

> "...In any case, we should shelve the loose assumption that the Laetoli footprints were made by Lucy's kind, *Australopithecus afarensis*."[102]

Homo Erectus

According to the evolutionary scenario, the transitional species between the australopithecines and true humans were given the designation, *Homo erectus*. Over time, a variety of fossil specimens have been put into this category. The most famous of all was Piltdown Man, which was exposed as a fraud many years ago.

Textbook Evolution

> "Some *Homo erectus* groups lived in caves and there is evidence that they also built crude wooden shelters. They successfully hunted large animals, butchered them using flint and bone tools, and cooked them over fires. *Homo erectus* living sites in China contain the remains of horses, bears, elephants, deer and

rhinoceroses."[103] (George B. Johnson, *Biology: Visualizing Life*, Holt, Rinehart, & Winston)

"(Figure 11.15 e) *Homo erectus* 1890, Java (now Indonesia) Physician Eugene Dubois discovered the top of a skull that could house a brain much larger than any ape's brain. It belonged to *Homo erectus* and was nicknamed "Java Man". Importance: Java Man's skull proved to be older than any other hominid fossils discovered up to that time."[104] (*Ibid.*)

A more recent edition of the Holt, Rinehart, and Winston *Biology* text informs us that Dubois was excited about this find because,

"First, the structure of the thighbone clearly indicated that the individual had long, straight legs and was an excellent, upright walker. Second, the size of the skull cap suggested a *very* large brain, about 900 cu. cm (54.9 cu. in.) --a great deal larger than any ape brain."[105] (George B. Johnson and Peter H. Raven, *Biology: Principles and Explorations*)

"(Figure 11.15 d) *Homo erectus* 1927-1938, Peking (now Beijing), China. Scientists unearthed a *Homo erectus* skull that closely resembled Java Man's skull (found in 1890). It was nicknamed "Peking Man." Fourteen other well preserved skulls, together with lower jaws and other bones, crude stone tools, and the ashes of campfires were also discovered. Importance: These finds showed that both Java Man and Peking Man belonged to the same species, *H. erectus*."[106] (George B. Johnson, *Biology: Visualizing Life*, Holt, Rinehart, & Winston)

"(Figure 11.15 f) *Homo erectus* 1984, Lake Turkana, Kenya. Scientists discovered a nearly complete skeleton of *Homo erectus*, dated at about 1.5 million years old (1 million years older than the skulls found in Java and Peking). Importance: The age of this fossil suggests that *Homo erectus* first evolved in Africa."[107] (*Ibid.*)

Science Versus

In the fossil record, with regard to primates, we really only have two categories: human skeletons and ape skeletons. The promoters of the theory of evolution have created two other categories which they justify on the basis of their evolutionary assumptions. The theorized creatures that are one step above the apes (ape man) are categorized as australopithecines. The theorized creatures that are one step below true humans (man apes) are categorized as *Homo erectus*. As we have shown throughout this chapter, after enough evidence is in, the australopithecines are shown to be simply specialized apes. Now we will show that the homo erectus fossils are coming to be understood as true human beings (homo sapiens).

In 1993, G. Philip Rightmire wrote the book, *The Evolution of Homo Erectus*. Michael Day, who reviewed the book and refers to *Homo erectus* as "the first true men," asked this important question, "Does *Homo Erectus* exist as a true species or should it be sunk into *Homo sapiens*?"[108]

G.W. Lasker, in his textbook, *Physical Anthropology*, answers the question as follows,

> "*Homo Erectus* is distinct from modern man (*Homo Sapiens*), but there is a tendency to exaggerate the differences...the range of variation of many features of *Homo Erectus* fall within that of modern man."[109]

Actually, the so-called *Homo erectus* fossils were those of true man, *Homo Sapien*. The researchers in the next three quotes are essentially agreeing with this statement.

William S. Laughlin, in an article in *Science* magazine states,

> "...When we find that significant differences have developed, over a short time span, between closely related and contiguous peoples, as in Alaska and Greenland, and when we consider the vast differences that exist between remote groups such as Eskimos and Bushmen, who are known to belong within the single species of *Homo sapiens*, it seems justifiable to conclude that *Sinanthropus* [*Homo erectus*] belongs within this same diverse species."[110]

Wolpoff, Xin Zhi, and Thorne, in the book, *The Origin of Modern Humans*, comment,

> "In our view, there are two alternatives. We should either admit that the *Homo Erectus/Homo sapiens* boundary is arbitrary and use nonmorphological (i.e. temporal) criteria for determining it, or *Homo Erectus* should be sunk [into *Homo sapiens*]."[111]

White and Brown, in their book, *The First Men*, report,

> "His bones were heavier and thicker than a modern man's and bigger bones required thicker muscles to move them. These skeletal differences, however, were not particularly noticeable. 'Below the neck,' one expert has noted, 'the differences between *Homo Erectus* and today's man could only be detected by an experienced anatomist.'"[112]

The following remarks by Earnst Hooton of Harvard illustrate the subjectivity in the business of representing our "ancestors,"

> "To attempt to restore the soft parts is an even more hazardous undertaking. The lips, the eyes, the ears, and the nasal tip, leave no clues on the underlying bony parts. You can with equal facility model on a Neanderthaloid skull the features of a chimpanzee or the lineaments of a philosopher. These alleged restorations of ancient types of man have very little if any scientific value and are likely only to mislead the public....So put not your trust in reconstructions."[113]

Dr. Bernard Wood, Professor of Human Origins at George Washington University, also reports that the pathway from animals to man is not as clear cut as the textbooks suggest,

> "...it is remarkable that the taxonomy and phylogenetic relationships of the earliest known representatives of our own genus, *Homo*, remain obscure. Advances in techniques for absolute dating and reassessments of the fossils themselves have rendered untenable a simple unilineal model of human evolution, in which *Homo habilis* succeeded the australopithecines and then evolved via

the *H. erectus* into *H. sapiens* - but no clear alternative consensus has yet emerged."[114]

Australopithecines Summary

When Professor Raymond Dart discovered the first so-called "australopithecine" in 1924, it was not met with gleeful acceptance in the scientific community. However, after the fall of Nebraska Man and Piltdown Man, it was elevated to a status of much importance. Nevertheless, as the years have gone by and more fossils have been found and more scientists have subjected these fossils to careful scrutiny, the high status of the australopithecines is beginning to crumble.

Charles Oxnard, Professor of Anatomy and Human Biology at the University of Western Australia, said of the australopithecines,

"The various australopithecines are, indeed, more different from both African apes and humans in most features than these latter are from each other. Part of the basis of this acceptance has been the fact that even opposing investigators have found these large differences as they too, used techniques and research designs that were less biased by prior notions as to what the fossils might have been."[115]

Oxnard's statistical analysis is certainly in disagreement with the idea that the australopithecines were bipedal tool makers. The three quotes below are his as well,

"Although most studies emphasize the similarity of the australopithecines to modern man, and suggest, therefore, that these creatures were bipedal tool-makers at least one form of which (*Australopithecus africanus -- 'Homo habilis,' 'Homo africanus*) was almost directly ancestral to man, a series of multivariate statistical studies of various postcranial fragments suggests other conclusions."[116]

"Multivariate studies of several anatomical regions, shoulder, pelvis, ankle, foot, elbow, and hand are now available for the australopithecines. These suggest that the common view, that these fossils are similar to modern man or that on those occasions when they depart from a similarity to man they resemble the African great apes, may be incorrect. Most of the fossil fragments are in fact uniquely different from both man and man's nearest living genetic relatives, the chimpanzee and gorilla....To the extent that resemblances exist with living forms, they tend to be with the orangutan."[117]

"Let us now return to our original problem: the australopithecine fossils....The new investigations suggest that the fossil fragments are usually uniquely different from any living form; when they do have similarities with living species, they are as often as not reminiscent of the orangutan."[118]

Actually, the best support for the ape/man candidates comes not from the paleontologist but from the artist, as Boyce Rensberger reports,

"Unfortunately, the vast majority of artist's conceptions are based more on imagination than on evidence. But a handful of expert natural-history artists begin with the fossil bones of a hominid and work from there....Much of the reconstruction, however, is guesswork. Bones say nothing about the fleshy parts of the nose, lips, or ears. Artists must create something between an ape and a human being; the older the specimen is said to be, the more apelike they make it....Hairiness is a matter of pure conjecture. The guesswork approach often leads to errors."[119]

The British anthropologist, Christopher Stringer, also acknowledges the challenge that each new discovery in the area of paleoanthropology provides,

"The study of human origins seems to be a field in which each discovery raises the debate to a more sophisticated level of uncertainty."[120]

Because the theory is wrong from the beginning, the discovery of new fossils generally adds more confusion than certainty. David Pilbeam of Yale believes firmly in evolution but readily admits to the problems,

"All this makes a much more complex picture of hominoid evolution than we once imagined. It no longer resembles a ladder but is, instead, more like a bush....Hominids evolved, as did many other mammal groups, with diverse and overlapping radiations. There is no clear-cut and inexorable pathway from ape to human being."[121]

Lord Solly Zuckerman of England has spent many years conducting research on the australopithecines. He remains unconvinced of the evidence that is supposed to support their ancestry to humans,

"The australopithecine skull is in fact so overwhelmingly simian as opposed to human...that the contrary proposition could be equated to an assertion that black is white."[122]

Pat Shipman of the department of Anthropology at Penn State comments,

"Interpreting the rich record of Miocene hominoids is fraught with problems because there are so few surviving examples. Today there are only three genera and four species of great apes, and these are confined to fast-vanishing tropical rainforest or woodland habitats. But there are about a dozen genera and even more species of larger-bodied hominoids known from the Miocene. This means that anthropologists are confronted with many ancient apes with no known living counterparts. Adding to the confusion are more smaller-bodied genera and species from the Miocene that appear to be related to monkeys, lesser apes, or have no apparent living relatives."[123]

She also stated in *Discover* magazine, regarding a new fossil find,

"What the new skull does, in a single stroke, is overturn all previous notions of early hominid evolution...The things we thought we understood reasonably well we don't."[124]

The ape-man possibilities of the australopithecines are diminishing, as reported by researchers, Cartmill, Pilbeam, and Isaac of Duke and Harvard,

"The australopithecines are rapidly shrinking back to the status of peculiarly specialized apes..."[125]

In fact, Dr. Oxnard concludes that australopithecines have no significant place in the evolution of humans,

> "...the australopithecines known over the last several decades...are now irrevocably removed from a place in the evolution of human bipedalism, possibly from a place in a group any closer to humans than to African apes and certainly from any place in the direct human lineage."[126]

Now and then even the high school textbook (Pearson/Prentice Hall) admits to the confusion in the effort to establish a line of descent from the apes to humans,

> "It is now clear that hominid evolution did not proceed by the simple, straight-line transformation of one species into another....Which hominids are true human ancestors? Which are just relatives? And how are all those species related to one another and to modern humans? At present no one can answer these questions."[127]

Photo-journalist and author of *Missing Links*, John Reader, acknowledges the prominent part of preconceptions in the business of clarifying the true apelike ancestors of mankind.

> "Because their antiquity could neither be proved nor disproved, the fossils themselves were the only evidence and, as was to be expected, interpretation of the evidence was decidedly colored by preconceptions concerning the theory of evolution in general"[128]

Anthropologist, Naoyuki Takahata, admits,

> "However, there are not enough fossil records to answer when, where, and how *H. sapiens* emerged."[129]

A little honesty is a refreshing thing. Here at the end of the human origins section, I think that David Pilbeam, one of the champions of Ramapithecus, provides us a sobering commentary on the whole history of digging up and proclaiming with such confidence the so called ape-like ancestors of man. This comment was made by Pilbeam in a review of Richard Leakey's book, *Origins*, in the journal, *American Scientist*,

"...perhaps generations of students of human evolution, including myself, have been flailing about in the dark; that our data base is too sparse, too slippery, for it to be able to mold our theories. Rather the theories are more statements about us and ideology than about the past."[130]

Summary

It has been incumbent upon the evolutionist to discover, in the fossil record, a series of intermediates between mankind and the animal world. They have chosen, as the candidates for this bridge, the primates. There has been an off the table argument among the key players as to whether man came from apes or ape-like creatures. Some have wanted to soft-pedal the idea that humans arose from the apes. For many, of course, this is an offensive idea. Nevertheless, we occasionally are informed that, like it or not, the apes are our ancestors. As previously quoted, Hooten states,

"If we are descended from apes our remote ancestors ought to look their part. You may not be willing to admit that you resemble an ape....But if that thousandth ancestor's forebearers become progressively more simian as you trace back the genealogical lines you will have to admit that somewhere in your family tree there squats an ape."[131]

Richard Lewontin, professor of Zoology at Harvard, said in an interview with Tom Bethell of *Harper's Magazine*,

"Look, I'm a person who says in this book [*Human Diversity,*1982] that we don't know anything about the ancestors of the human species. All the fossils which have been dug up and are claimed to be ancestors- we haven't the faintest idea whether they are ancestors....All you've got is Homo sapiens there, you've got that fossil there, you've got another fossil there...and it's up to you to draw the lines. Because there are no lines."[132]

Roger Lewin, in his book, *Bones of Contention*, reports that,

"Richard and his parents, Louis and Mary, have held to a view of human origins for nearly half a century now that the line of true man, the line of *Homo*--large brain, tool making and so on--has a separate ancestry that goes back millions of millions of years. And the ape-man, *Australopithecus*, has nothing to do with human ancestry."[133]

Author Robert Martin, in his book, *Primate Origins and Evolution*, provides information that should be in the high school textbooks but is not,

> "It should be noted at the outset that substantial fossil remains are known for all of the species listed below (a quite unusual situation with respect to the primate fossil record generally), but that there is virtually no fossil evidence relating to human evolution, other than a few fragments of dubious affinities, before about 3.8 my ago. The preceding period of human evolution therefore remains a complete mystery and an unfortunate major gap exists whatever view one takes of the time of divergence of hominids and great apes."[134]

Adler and Carey, in their *Newsweek* article, reveal the frustration encountered by those who are trying to establish man's ape-like ancestors,

> "The missing link between man and the apes, whose absence has comforted religious fundamentalists since the days of Darwin, is merely the most glamorous of a whole hierarchy of phantom creatures....The more scientists have searched for the transitional forms that lie between species, the more they have been frustrated." [135]

Probably, most people do not know of Louis Leakey's final attitude with regard to Darwin's perspective of human evolution,

> "Dr. Leaky bases his repudiation of Darwin on the results of his long search in East Africa for the remains of the original man. The generally accepted post-Darwin view is that man developed from the baboon 3 to 5 million years ago. But Leakey has found no evidence of a spurt in development at that time."[136]

Lord Solly Zuckerman's antagonism toward the idea that the australopithecines are man's ancestor is referenced by Lewin in his book,

> "His Lordship's scorn for the level of competence he sees displayed by paleoanthropologists is legendary, exceeded only by the force of his dismissal of the australopithecines as having anything at all to do with human evolution. 'They are just bloody apes,' he is reputed to have observed on examining the australopithecine remains in South Africa....Zuckerman had become extremely powerful in British science, being an adviser to the government up to the highest level....While at Oxford and then Birmingham universities, he had vigorously pursued a metrical and statistical approach to studying the anatomy of fossil hominids. No secure inference could be drawn without such an analytical approach, he urged, and it was on this basis that he underpinned his lifelong rejection of the australopithecines as human ancestors."[137]

F. Clark Howell, an evolutionary biologist at Berkeley, states,

> "There is no encompassing theory of evolution...there never really has been."[138]

The search for the common ancestor for chimpanzees and humans remains, to this day, for the evolutionists, a perplexing and aggravating mystery. Jennifer Viegas reveals this in an article in *Discovery News*,

> "The last common ancestor of chimpanzees and humans remains a holy grail in science."[139]

Subjectivism in interpreting the meager fossil evidence for human evolution has often played an inordinately large part in supporting the theory of evolution. A variety of science researchers admit to this, as seen in the quotes below, beginning with Roger Lewin,

> "In fact, 'virtually all our theories about human origins were relatively unconstrained by fossil data,' observes David Pilbeam. 'The theories are ... fossil-free or in some cases even fossil-proof.' This shocking statement simply means that there is and always has been far more fleshing out of the course and cause of human evolution than can fully be justified by the scrappy skeleton provided by the fossils. As a result, he continues, 'our theories have often said far more about the theorists than they have about what actually happened.'"[140]

Niles Eldredge concludes,

"And it has been the paleontologist--my own breed--who have been most responsible for letting ideas dominate reality: ...We paleontologists have said that the history of life supports that interpretation [gradual adaptive change], all the while knowing that it does not."[141]

"I simply thought that the time had come to take the fossil record -- the patterns of stability and change -- a bit more literally than had traditionally been the case."[142]

If he would have taken it truly literally, he should have concluded that the theory of evolution has no support from the fossil record and therefore there is no historical, documentary support for this theory.

As referenced by John Reader, preconceptions also play a leading role in this battle,

"The entire hominid collection known today would barely cover a billiard table, ... the collection is so tantalizingly incomplete, and the specimens themselves often so fragmented and inconclusive, that more can be said about what is missing than about what is present. ...but ever since Darwin's work inspired the notion that fossils linking modern man and extinct ancestor would provide the most convincing proof of human evolution, preconceptions have led evidence by the nose in the study of fossil man."[143]

Over and again, subjectivism weighs heavily in the modern science of paleoanthropology, as acknowledged by David Pilbeam of Yale University,

"My reservations concern not so much this book but the whole subject and methodology of paleoanthropology. But introductory books-- or book review-- are hardly the place to argue that perhaps generations of students of human evolution, including myself, have been flailing about in the dark; that our data base is too sparse, too slippery, for it to be able to mold our theories. Rather, the theories are more statements about us and ideology than about the past. Paleoanthropology reveals more about how humans view themselves than it does about how humans came about."[144]

He further comments,

"I am also aware of the fact that, at least in my own subject of paleoanthropology, 'theory'--heavily influenced by implicit ideas--almost always dominates 'data.'...Ideas that are totally unrelated to actual fossils have dominated theory building, which in turn strongly influences the way fossils are interpreted."[145]

J.S. Jones, reporting in *Nature* magazine, lets the cat out of the bag with regard to the kind of science that is involved in the discipline of paleoanthropology,

"Paleoanthropologists seem to make up for a lack of fossils with an excess of fury, and this must now be the only science in which it is still possible to become famous just for having an opinion. As one cynic says, in human paleontology the consensus depends on who shouts loudest."[146]

Further, as Ronald West, assistant professor of paleobiology at Kansas State University admits, the employment of circular reasoning also undermines the scientific aspect of the theory of evolution,

"Contrary to what most scientists write, the fossil record does not support the Darwinian theory of evolution because it is this theory (there are several) which we use to interpret the fossil record. By doing so we are guilty of circular reasoning if we then say the fossil record supports this theory."[147]

Regarding the clever pictures of man evolving upward from apes, Professor of Human Origins, Bernard Wood, of George Washington University, clarifies,

"There is a popular image of human evolution that you'll find all over the place...On the left of the picture there's an ape...On the right, a man...Between the two is a succession of figures that become ever more like humans...Our progress from ape to human looks so smooth, so tidy. It's such a beguiling image that even the experts are loath to let it go. But it is an illusion."[148]

In summary, it seems that when the facts are all collected and the fossils are examined, the picture that emerges does not differ significantly from the ancient Genesis account. Work done by Michael Hammer at the University of Arizona and reported on by *US News and World Report,* is certainly supportive of the Genesis account,

> "We are finding that humans have very, very shallow genetic roots which go back very recently to one ancestor... that indicates that there was an origin in a specific location on the globe and then it spread out from there... researchers suggest that virtually all modern men- 99.9% of them, says one scientist- are closely related genetically and share genes with one male ancestor, dubbed 'Y-chromosome Adam.'"[149]

On the contrary, the theory of evolution, which has put so much stock in the fossil record as a means of providing scientific support for this highly imaginative idea, has been consistently disappointed by the paleontological research. For instance, in an article in *Newsweek* magazine, Jerry Adler, *Newsweek* contributing editor, and John Carey, write,

> "The missing link between man and apes...is merely the most glamorous of a whole hierarchy of phantom creatures. In the fossil record, missing links are the rule...The more scientists have searched for the transitional forms between species, the more they have been frustrated."[150]

Science is what you see. What we see in the real world, as opposed to the imaginary and the theoretical, are the following kinds of groups of animals: invertebrates, fishes, amphibians, reptiles, mammals, birds, primates (apes and monkeys) and humans. These groups are easily associated with the groups referenced in the book of Genesis as being created by God, and as being created "after their kind." If evolution was in fact a real and historical process, there should be a significant number of transitional species between these various groups for us to observe in the present. Since there are none whatsoever, this should make anyone who is being objective about this theory, highly suspicious of the theory of evolution. Inasmuch, as there are no transitional organisms between one kind and another in our present world, the evolutionist, in order to make his theory credible to those who are willing to believe it, has had to propose that this amazing transition from one animal to another must have taken place slowly and gradually over millions and millions of years of time. Thus, the evolutionists are stuck with the fossil record as providing the only means by which they can, hopefully, document this proposed history. As Dunbar of Yale states,

> "Although the comparative study of living plants and animals may give very convincing circumstantial evidence, fossils provide the only historical, documentary evidence that life has evolved from simpler to more and more complex forms."[151]

So, Dunbar seems to be making the point to his students and readers that comparing one group to another is inadequate as reliable evidence for evolution. There must be fossils that show them to be the remains of organisms that were actually transitional from one group to another (such as from invertebrates to fish). This means, of course, that if there are no such transitional fossils discovered, then there is no historical documentary evidence for evolution.

As many of the readers of this text must surely realize by now, this embarrassing lack of transitional fossils has forced many authorities within the evolutionary camp to propose and escape from this unpleasant situation. The terminology associated with this life-saving escape, as mentioned earlier, has been called punctuated equilibrianism. Instead of admitting that the lack of transitional fossils very clearly supports special creation and the Genesis record of origins, they come up with the novel idea that this lack actually proves evolution. How can the lack of transitional fossils prove evolution? According to the evolutionists, applying a most unique form of logic, they contend that this fossil deficiency proves that evolution took place rapidly in spurts. They argue that the lack of transitions shows that during the theorized millions of years that it took for one group to be changed into another (such as reptile to birds) there was, unfortunately for the theory of evolution, no sedimentation being laid down anywhere in the world which might have preserved, as fossils, these theorized transitional types. This outlandish idea becomes, perhaps, the only theory in the history of science to be based upon negative evidence. Nevertheless, the loyal evolutionist is quite comfortable with such a proposal. As Stanley of Johns Hopkins University relates,

"Established species are evolving so slowly that major transitions between genera and higher taxa must be occurring within small, rapidly evolving populations that leave no legible fossil record."[152]

And finally, as a result of the new and improved theory of evolution, punctuated equalibrianism, Mark Ridley of Oxford indicates that, for all intents and purposes, the fossil record has become irrelevant to evolution. His words are (as previously referenced), "In any case, no real evolutionist...uses the fossil record as evidence in favor of the theory of evolution as opposed to special creation..."

This further means that, if no such transitional fossils are discovered and evolutionists continue to maintain their confidence in this theory, they are inadvertently admitting that their theory is true by definition and that "historical documentary evidence" is unnecessary.

As we look at the two different models of origins, the Creation model and the Evolution model, it should become abundantly clear that the fossil evidence is not in conflict with the Creation model, but it most definitely is in conflict with the Evolution model.

Chapter Two End Notes

[1] Earnest Mayr, "Interview," *Omni,* March/April 1988, p. 46
[2] Earnst A. Hooton, *Up From the Ape,* McMillan, 1931, p. 289
[3] Sylvia S. Mader, *Biology,* McGraw Hill, 2001, p. 341
[4] Kenneth R. Miller & Joseph Levine, *Biology*, Pearson/Prentice Hall, 2004, p. 837
[5] Eric Strauss & Marylin Lisowski, *Biology The Web of Life*, Second Edition, Scott-Foresman-Addison Wesley, p.280-281
[6] A.J. Kelso, *Physical Anthropology*, Lippincott, second edition, 1974, p. 142
[7] A.S. Romer, Vertebrate Paleontology, 3rd edition, University of Chicago Press, 1966, p.218
[8] C.B.G. Campbell, *Science*, vol. 153, 1966, p. 436
[9] Anthea Maton, *Evolution: Change Over Time*, Prentice Hall Science, 1993, p. 80
[10] *Ibid.*
[11] Alton Biggs, Chris Kapicka, Linda Lundgren, *Biology*, *The Dynamics of Life*, Glencoe, McMillian/ McGraw-Hill,1995, p. 456
[12] *Ibid.*, p. 459
[13] Anthea Maton, *Evolution: Change Over Time*, Prentice Hall Science, 1993, p. 80
[14] E.L. Simons, *Annals New York Academy of Sciences*, Vol. 167, 1969, p. 319
[15] Anthea Maton, *Evolution: Change Over Time*, Prentice Hall Science, 1993, p. 81
[16] Alton Biggs, Chris Kapicka, Linda Lundgren, *Biology*, *The Dynamics of Life*, Glencoe,McMillian/ McGraw-Hill, 1995, p. 459
[17] Alton Biggs, et. al., *Biology: The Dynamics of Life*, Glencoe Science, *National Geographic*, McGraw-Hill/Glencoe, 2006, p. 426
[18] William Howells, "Mankind in the Making," Doubleday and Co., 1959, p. 102
[19] A.J. Kelso, *Physical Anthropology*, Lippincott Williams & Wilkins, 1974, p. 151
[20] Robert D. Martin, *Primate Origins and Evolution*, Princeton University Press, 1990, p. 69
[21] Anthea Maton, *Evolution: Change Over Time*, Prentice Hall Science, 1993, p. 80
[22] Alton Biggs, Chris Kapicka, Linda Lundgren, *Biology*, *The Dynamics of Life*, Glencoe, McMillian/ McGraw-Hill,1995, p. 459
[23] Alton Biggs, et. al., *Biology: The Dynamics of Life*, Glencoe Science, *National Geographic*, McGraw-Hill/Glencoe, 2006, p. 426
[24] A.J. Kelso, *Physical Anthropology*, Lippincott Williams & Wilkins, 1974, p. 142
[25] Robert D. Martin, *Primate Originsmand Evolution*, Princeton University Press, 1990, p. 66
[26] William L. Straus, Jr., A.L. Kroeber, ed. *Anthropology Today*, University of Chicago Press, 1953, p. 77
[27] Alton Biggs, Chris Kapicka, Linda Lundgren, *Biology*, *The Dynamics of Life*, Glencoe, McMillian/ McGraw-Hill, 1995, p. 462
[28] *Ibid.*
[29] Eric Strauss & Marylin Lisowski, *Biology: The Web of Life*, second edition, Scott Foresman-Addison Wesley, 2000, p. 281
[30] Lyall Watson, *Science Digest*, vol. 90, May 1982, p. 44
[31] David Pilbeam, "Rearranging Our Family Tree," *Human Nature*, June 1978, p. 43
[32] Donald Johanson, *Lucy: The Beginnings of Humankind*, Simon & Schuster, 1981, p. 363
[33] G.H.R. von Koenigswald, *Meeting Prehistoric Man*, Harper and Bros., 1956, p. 199
[34] Henry Gee, *Nature*, July 12, 2001, p. 131
[35] William King Gregory, "The Origin of Man from a Brachiating Anthropoid," *Science*, Vol. 71, 1930, p. 650
[36] Elwyn Simons, "The Phyletic Position of *Ramapithecus*," *Postilla*, Yale Peabody Museum, 1961, p. 373
[37] Elwyn Simons, "On the Mandible of Ramapithecus" *Anthropology*, Vol. 54, 1964
[38] Peter Andrews, "Hominid Evolution," *Nature*, vol. 295, 1982, p. 186
[39] Elwyn Simons, Interview with the author, Duke University, 25 September 1985 (as referenced in *Bones of Contention*, p. 124)

[40] David Pilbeam, "Rethinking Human Origins," in *Discovery*, vol. 13, 1978, p. 9
[41] David Pilbeam, *American Scientist*, Vol. 66, May/June 1978, p. 379
[42] Roger Lewin, *Bones of Contention*, A Touchstone Book, Simon & Schuster, 1988, p. 87
[43] John H. Postlethwait & Janet L. Hopson, *Modern Biology*, Holt, Rinehart, & Winston, 2006, p. 878
[44] Alton Biggs, Whitney Crispen Hagins, William G. Holliday, Chris L. Kapicka, Linda Lundgren, Ann Haley MacKenzie, William D. Rogers, Marion B. Sewer, Dinah Zike, *Biology,* Glencoe, 2007, p. 464
[45] George B. Johnson, *Biology: Visualizing Life*, Holt, Rinehart & Winston, 1994, p. 229
[46] *Ibid.*
[47] John H. Postlethwait & Janet L. Hopson, *Modern Biology*, Holt, Rinehart, & Winston, 2006, p. 878
[48] Charles Oxnard, "Human Fossils: New View of Old Bones," *American Biology Teacher*, May 1979
[49] *Ibid.*
[50] Charles Oxnard, *Uniqueness and Diversity in Human Evolution: Morphometric Studies of the Australopithecines*, University of Chicago Press, 1975
[51] J.B. Birdsell, *Human Evolution,* 2nd Ed., Rand McNally College Publishing Co., 1975 p. 60
[52] Kenneth P. Oakley, *Dating the Emergence of Man,* Bobbs-Merrill, 1962, p. 34
[53] David Lambert and the Diagram Group, *A Field Guide to Early Man*, Facts on File, 1987, p. 101
[54] George B. Johnson, Holt, Rinehart, & Winston, *Holt Biology: Visualizing Life*, 1994, p. 229
[55] George B. Johnson and Peter H. Raven, *Biology: Principles and Explorations*, Holt, Rinehart and Winston, p. 301
[56] Roger Lewin, *Bones of Contention*, A Touchstone Book, Simon & Schuster, 1988, p. 141
[57] *Ibid.*, p. 140
[58] Louis Leakey, "Finding the World's Earliest Man," in *National Geographic*, September 1961, p. 433
[59] Louis Leakey, "A New Fossil Skull from Olduvai," *Nature*, vol. 184, 1959, pp. 491-493
[60] Roger Lewin, *Bones of Contention,* A Touchstone Book, Simon & Schuster Inc., 1988, p. 150
[61] David Lambert and the Diagram Group, *The Field Guide to Early Man*, Facts on File, 1987, p. 102
[62] *Ibid.*
[63] *Ibid.* p. 104
[64] Boyce Rensberger, *Science* 84, April, p. 34
[65] *Modern Biology*, Holt, Rinehart, & Winston, 2002, p. 325
[66] Phillip Tobias, Letter to *Discovery*, August 1964, pp. 49-50
[67] Louis Leakey, Letter to *Discovery*, October 1964, p. 68
[68] Roger Lewin, *Bones of Contention*, A Touchstone Book, Simon & Schuster Inc., 1988, p. 145
[69] George B. Johnson, *Biology: Visualizing Life*, Holt, Rinehart, & Winston, 1994, p. 232
[70] George B. Johnson & Peter H. Raven, *Biology: Principles and Explorations*, Holt, Rinehart, and Winston, 1996, p. 303
[71] Eric Strauss & Marylin Lisowski, *Biology: The Web of Life*, second edition, Scott Foresman-Addison Wesley, 2000, p. 288
[72] *Modern Biology*, Holt, Rinehart, & Winston, 2002, p. 327
[73] Alton Biggs, et. al., *Biology: The Dynamics of Life*, Glencoe Science, *National Geographic*, McGraw-Hill/Glencoe, 2006, p. 879
[74] Henry McHenry, *Science*, vol. 236, p. 1061
[75] *Ibid.*
[76] Donald Johanson, *Science*, Vol 236, p. 1061
[77] Henry McHenry, Op Cit.
[78] Henry McHenry, Op Cit.
[79] Ian Tattersall, *Evolutionary Anthropology*, 1(1): 34-36, 1992
[80] C. Loring Brace quoted in *Lucy: The Beginnings of Humankind,* by Donald Johanson & Maitland Edey, Simon & Schuster, 1981, p. 97
[81] George Johnson and Peter Raven, *Biology Principles and Explorations,* Holt, Rinehart and Winston, 1996, p. 301
[82] *Modern Biology*, Holt, Rinehart, & Winston, 2002, p. 324
[83] Kenneth R. Miller & Joseph Levine, *Biology*, Pearson/Prentice Hall, 2008, p. 837

[84] *(Ibid.)*
[85] William Fix, *The Bone Peddlers*, Scribner, 1984, p. 73
[86] *Science News*, Vol. 131, p. 58
[87] Referring to comments made by Richard Leakey in *The Weekend Australian*, 7-8 May 1983, Magazine, p. 3
[88] A. L. Zihlman, "Pygmy Chimps, People, and the Pundits," *New Scientist*, Vol. 104, No. 1430, Nov. 1984, p. 39
[89] David Lambert, *A Field Guide to Early Man*, Facts on File, 1987, p. 98
[90] David Lambert, *A Field Guide to Early Man*, Facts on File, 1987, p. 99
[91] *Nature*, 376, 8/17/1995, p. 556
[92] William Howells, *Getting Here*, The Compass Press, 1993, p. 79
[93] Adrienne L. Zihlman, *Science News*, vol. 123, Feb. 5, 1983, p. 89
[94] George B. Johnson, *Biology: Visualizing Life*, Holt, Rinehart, & Winston, 1994, p 230
[95] Brian G. Richmond and David S. Strait, March 23, 2000, "Evidence that humans evolved from a knuckle-walking ancestor," *Nature,* 404:382-385
[96] D. Schmitt, *Journal of Experimental Biology,* 2003, 206:1437-1448
[97] *Modern Biology*, Holt, Rinehart, & Winston, 2002, p. 323
[98] George B. Johnson, *Biology: Visualizing Life*, Holt, Rinehart, & Winston, 1994, p. 229
[99] Kenneth R. Miller & Joseph Levine, *Biology*, Pearson/Prentice Hall, 2008, p. 837
[100] *TIME*, Nov. 10, 1975, p. 93
[101] Russell H. Tuttle, "The Pitted Pattern of Laetoli Feet," *Natural History*, vol. 99, 3/90, p. 64
[102] *Ibid.*
[103] George B. Johnson, *Biology: Visualizing Life*, Holt, Rinehart, & Winston, 1994, p. 233
[104] *Ibid.*
[105] George B. Johnson and Peter H. Raven, *Biology: Principles and Explorations*, Holt, Rinehart, and Winston, 1996, p. 303
[106] George B. Johnson, Biology: Visualizing Life, Holt, Rinehart, & Winston, 1994, p. 233
[107] *Ibid.*
[108] Michael Day, *"Homo* Turmoil", *Nature*, 348, 20/27 December, 1990, p. 688
[109] G.W. Lasker, *Physical Anthropology*, Holt, Rinehart, and Winston, 1973, p. 284
[110] W.S. Laughlin, "Eskimos and Aleuts: Their Origin and Evolution," *Science* 142, November 8, 1963, p. 644
[111] M.H. Wolpoff, Wu Xin Zhi, and A.G. Thorne, "Modern *Homo sapiens* Origins: A General Theory of Hominid Evolution Involving the Fossil Evidence From East Asia," *The Origins of Modern Humans*, Fred H. Smith and Frank Spencer, eds., Alan R. Liss, Inc., 1984, p. 465-66, (bracketed for clarity)
[112] Edmund White and Dale Brown, *The First Men*, Time-Life Books, 1973, p. 14
[113] Earnst A. Hooton, *Up From the Ape,* McMillan, 1931, p. 332
[114] Bernard Wood, *Nature* 355, 783, 1992
[115] Charles Oxnard, *Fossils, Teeth and Sex--New perspective on Human Evolution*, University of Washington Press, 1987, p. 227
[116] Charles E. Oxnard, "The Place of the Australopithecines in Human Evolution: Grounds for Doubt?" *Nature*, Vol. 258, December 1975, p. 389
[117] Charles E. Oxnard, *University of Chicago Magazine*, Winter 1974, p. 11
[118] Charles E. Oxnard, "Human Fossils: New View of Old Bones," *American Biology Teacher*, Vol. 41, May 5, 1979, p. 273
[119] Boyce Rensberger, *Science Digest*, 1981
[120] Christopher B. Stringer, "The Legacy of Homo Sapiens," *Scientific American,* Vol. 268, May 1993, p. 138
[121] David Pilbeam, "Rearranging Our Family Tree," *Human Nature*, June 1978, pp. 44, 45
[122] Lord Solly Zuckerman, *Beyond the Ivory Tower: the frontiers of public and private sciences,* Taplinger Publishing Company, 1971, p. 78
[123] Pat Shipman, "Where Have all the Primates Gone?" *New Scientist*, Vol. 134, June 13, 1992, p. 16
[124] Pat Shipman, *Discover* magazine, September 1986

[125] Matt Cartmill, David Pilbeam, Glynn Isaac, "One Hundred Years of Paleoanthropology," *American Scientist* 74:419, July-August, 1986, p 419
[126] Charles Oxnard, *The Order of Man*, Yale University Press, p. 332
[127] Kenneth R. Miller & Joseph Levine, *Biology*, Pearson/Prentice Hall, 2004, p. 838
[128] John Reader, *Missing Links*, Little Brown & Co, First American Edition edition, 1981, p. 25
[129] Takahata, *Annual Review of Ecology and Systematics*, 1995
[130] David Pilbeam, *American Scientist*, Vol. 66, May/June 1978, p. 379
[131] Earnst A. Hooton, *Up From the Ape,* McMillan, 1931, p. 289
[132] Tom Bethell interviewing Richard C. Lewontin, "Agnostic Evolutionists" *Harper's Magazine,* Feb. 1985, pp. 60-61
[133] Roger Lewin, *Bones of Contention,* A Touchstone Book, Simon & Schuster Inc., 1988, pp. 17-18
[134] Robert D. Martin, *Primate Origins and Evolution*, Princeton University Press, 1990, p. 69
[135] Jerry Adler & John Carey, "Is Man a Subtle Accident?" *Newsweek,* November 3, 1980, pp. 54-55
[136] *Chicago American*, 1/25, 1967
[137] Roger Lewin, *Bones of Contention,* A Touchstone Book, Simon & Schuster Inc., 1988, p. 164, 165
[138] F. Clark Howell, "Thoughts on the Study and Interpretation of the Human Fossil Record," *Contemporary Issues in Human Evolution,* W. Eric Meikle, F. Clark Howell, and Nina G., editors, California Academy of Sciences, 1996, p. 21
[139] Jennifer Viegas, "The Human Family Tree," *Discovery News*, Posted on discovery.com, accessed November 11, 2010
[140] Roger Lewin, *Bones of Contention*, Simon & Schuster Inc., 1987, p. 43
[141] Niles Eldridge, *Time Frames*, 1986, p. 144
[142] Niles Eldredge, "Evolution in real time" pp. 96-97 or p. 60
[143] John Reader, "Whatever Happened to Zinjanthropus?" *New Scientist*, 26 March 1981, p. 802
[144] David Pilbeam, review of *Origins*, by Richard E. Leakey and Roger Lewin (Dutton, 1977, p. 264) *American Scientist*, Vol. 66 May/June 1978, pp. 378-9
[145] David Pilbeam, "Rethinking Human Origins," *Discovery*, vol. 13, 1978, pp. 8-9
[146] J.S. Jones, "A Thousand and One Eves," *Nature,* Vol. 345, May 31, 1990, p. 395
[147] Ronald R. West, "Paleoecology and Uniformitarianism," *Compass*, vol. 45, May 1968, p. 216
[148] Bernard Wood, *New Scientist*, October 10, 2002
[149] Michael Hammer, *US News and World Report*, Dec. 4, 1995
[150] Jerry Adler and John Carey, "Is Man a Subtle Accident?," *Newsweek*, Nov. 2, 1980
[151] C.O. Dunbar, *Historical Geology,* 2nd ed., John Wiley & Sons, 1960, p. 47
[152] Steven M. Stanley, "Macroevolution and the Fossil Record," *Evolution*, 36, No. 3, 1982, p. 460

Made in the USA
Monee, IL
08 November 2026

67380802R00083